Enzyme Biotechnology

Enzyme Biotechnology

Annabelle Pearson

Callisto Reference

www.callistoreference.com

Callisto Reference,
118-35 Queens Blvd., Suite 400,
Forest Hills, NY 11375, USA

Visit us on the World Wide Web at:
www.callistoreference.com

ISBN: 978-1-64116-544-0 (Hardback)

Cataloging-in-Publication Data

Enzyme biotechnology / Annabelle Pearson.
p. cm.
Includes bibliographical references and index.
ISBN 978-1-64116-544-0
1. Enzymes--Textbooks. 2. Enzymes--Biotechnology--Textbooks. 3. Enzymology--Textbooks.
4. Biochemistry--Textbooks. 5. Biotechnology--Textbooks. I. Pearson, Annabelle.
QP601 .E59 2022
572.7--dc23

TABLE OF CONTENTS

PREFACE

The purpose of this book is to help students understand the fundamental concepts of this discipline. It is designed to motivate students to learn and prosper. I am grateful for the support of my colleagues. I would also like to acknowledge the encouragement of my family.

The macromolecular biological catalysts that accelerate the chemical reactions are known as enzymes. The molecules upon which enzymes react are known as substrates. These are converted into different molecules known as products. Enzymes are produced using recombinant expression in selected host microorganisms, recovery, fermentation and formulation. Enzymes are used in a variety of industries and fields such as biofuel industry, brewing industry and chemical industry. The largest user of enzymes is the detergent industry. They are utilized to remove soiling, protect garments and increase softness. Enzymes are also used in the textile, pulp and paper, leather, and animal feed industries. This book provides significant information of this discipline to help develop a good understanding of enzyme biotechnology and related fields. It aims to shed light on some of the unexplored aspects of this field. This book will serve as a valuable source of reference for graduate and post graduate students.

A foreword for all the chapters is provided below:

Chapter – Understanding Enzymes

The macromolecular biological catalysts which are involved in the acceleration of chemical reactions are called enzymes. The small organic molecules which are involved in the transportation of chemical groups from one enzyme to the other are known as coenzymes. This is an introductory chapter which will introduce briefly all the significant aspects of enzymes and coenzymes.

Chapter – Types of Enzymes

Some of the different types of enzymes are lipase, amylase, trypsin and lactase. Lipases are enzymes which are responsible for the catalysis of the hydrolysis of fats. They perform an integral role in digestion as well as the processing of dietary lipids. The topics elaborated in this chapter will help in gaining a better perspective about these diverse types of enzymes.

Chapter – Enzyme Kinetics

The branch of science which deals with the study of the chemical reactions which are catalyzed by enzymes is known as enzyme kinetics. The increase which happens due to the presence of an enzyme in the rate of a process is known as enzyme catalysis. The chapter closely examines these key concepts of enzyme kinetics to provide an extensive understanding of the subject.

Chapter - Enzyme Inhibition

The process through which the catalytic efficiency of an enzyme is decreased is known as enzyme inhibition. The molecules which cause inhibition in enzymes are known as inhibitors. A few types of inhibitors are uncompetitive inhibitors and targeted covalent inhibitors. All the diverse processes related to enzyme inhibition have been carefully analyzed in this chapter.

Chapter - Applications of Enzymes

Enzymes find applications in a variety of industries as extremely specific catalysts. Some of the industries where they are used are biofuel industry, dairy industry, brewing industry and paper industry. These diverse applications of enzymes have been thoroughly discussed in this chapter.

Annabelle Pearson

Understanding Enzymes

CHAPTER 1

The macromolecular biological catalysts which are involved in the acceleration of chemical reactions are called enzymes. The small organic molecules which are involved in the transportation of chemical groups from one enzyme to the other are known as co-enzymes. This is an introductory chapter which will introduce briefly all the significant aspects of enzymes and coenzymes.

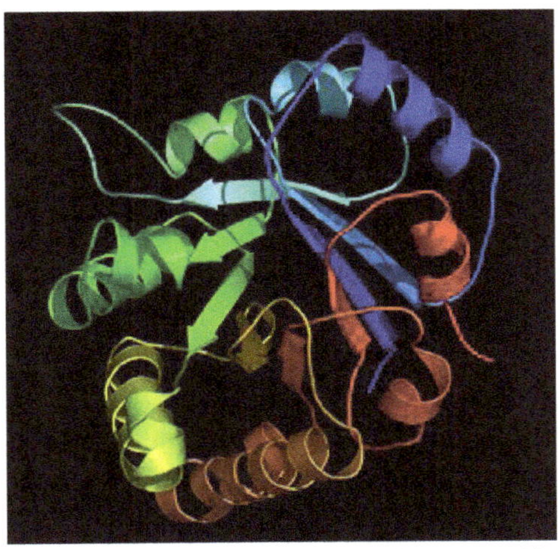

Ribbon diagram of the enzyme TIM. Each enzyme has a specific three-dimensional structure that determines its function.

An enzyme is a biological catalyst that regulates the rate of a chemical reaction in a living organism. Most enzymes are proteins, though certain nucleic acids, called ribozymes, are also capable of catalytic activity.

Enzymes are essential to sustain life because most chemical reactions in biological cells, such as the digestion of food, would occur too slowly or would lead to different products without the activity of enzymes. Most inherited human diseases result from a genetic mutation, overproduction, or deficiency of a single critical enzyme. For example, lactose intolerance, the inability to digest significant amounts of lactose, which is the major sugar found in milk, is caused by a shortage of the enzyme lactase.

For an enzyme to be functional, it must fold into a precise three-dimensional shape. How such a complex folding can take place remains a mystery. A small chain of 150 amino acids making up an enzyme has an extraordinary number of possible folding configurations: if it tested 1012 different configurations every second, it would take about 1026 years to find the right one. Yet, a denatured enzyme can refold within fractions of a second and then precisely react in a chemical reaction. To some, it suggests that quantum effects are at work even at the large distances (by atomic standards) spanned by a protein molecule. At least, it demonstrates a stunning complexity and harmony in the universe.

While all enzymes have a biological role, some enzymes are also used commercially. For instance, many household cleaners use enzymes to speed up the breakdown of protein or starch stains on clothes.

Like all catalysts, enzymes work to lower the activation energy of a reaction, or the initial energy input necessary for most chemical reactions to occur. Heat cannot be added to a living system, so enzymes provide an alternate pathway: they bond with a substrate (the substance involved in the chemical reaction) to form a "transition state," an unstable intermediate complex that requires less energy for the reaction to proceed. Like any catalyst, the enzyme remains unaltered by the completed reaction and can therefore continue to interact with substrates. Enzymes may speed up reactions by a factor of many millions.

Enzymes can be affected by molecules that increase their activity (activators) or decrease their activity (inhibitors). Many drugs act by inhibiting enzymes. Aspirin works by inhibiting COX-1 and COX-2, the enzymes that produce prostaglandin, a hormonal messenger that signals inflammation. By inhibiting the activity of these enzymes, aspirin suppresses our experience of pain and inflammation.

The Structure of Enzymes

Enzyme structure is important because it determines the enzyme's particular function in the body. Enzymes (and other proteins) are composed of amino acid chains called polypeptide chains. The linear sequence of amino acids determines the characteristic folding of the chains into a three-dimensional structure. An enzyme might contain only one polypeptide chain, typically linking one hundred or more amino acids, or it might consist of several polypeptide chains that act together as a unit.

Most enzymes are larger than the substrates on which they act. Only a very small portion of the enzyme, approximately ten amino acids, comes into direct contact with the substrate(s). This region, where the binding of the substrate(s) and the reaction occur, is known as the active site of the enzyme.

Specificity

Enzymes are usually specific, or unique, to the reactions they catalyze and the substrates that are involved in these reactions. An enzyme combines with its substrate(s) to form a short-lived enzyme-substrate complex. There are two models to explain how the binding of enzyme and substrate occurs: the "lock and key" model and induced fit.

Lock and Key Model

To account for the specificity of enzymes, Emil Fischer proposed that the enzyme had a particular shape into which the substrate(s) fit exactly. This model of exact fit, introduced in the 1890s, is often referred to as the "lock and key" model, because the enzyme binding to a substrate is analogous to the specific fit of a lock into a key.

Induced Fit Model

In 1958, Daniel Koshland suggested a modification to the "lock and key" model. Unlike keys, enzymes are rather flexible structures. The active site of an enzyme can be modified as the substrate interacts with the enzyme, creating an "induced fit" between enzyme and substrate. The amino acids side chains that make up the active site are molded into a precise shape, which enables the enzyme to perform its catalytic function. In some cases, the substrate molecule changes shape slightly as it enters the active site.

Enzyme Cofactors

Some enzymes do not need any additional components to exhibit full activity. However, others require non-protein molecules to be bound to the complex for efficient activity. Cofactors can be either inorganic (e.g., metal ions and iron-sulfur clusters) or organic compounds, which are also known as coenzymes.

Most cofactors are not covalently bound to an enzyme, but are closely associated. However, some cofactors known as prosthetic groups are tightly bound to the enzyme through covalent bonds.

Most cofactors are either regenerated or chemically unchanged at the end of the reactions. Many cofactors are vitamin-derivatives. They serve as carriers during the reaction to transfer electrons, atoms, or functional groups from an enzyme to a substrate. Common examples include NAD and NADP, which are involved in electron transfer, and coenzyme A, which is involved in the transfer of acetyl groups.

How Enzymes Catalyze Reactions

A reaction catalyzed by enzymes must be spontaneous; that is, having a natural tendency to occur without needing an external push. (Thermodynamically speaking, the reaction must contain a net negative Gibbs free energy.) In other words, the reaction would run in the same direction without the enzyme, but would occur at a significantly slower rate. For example, the breakdown of food particles such as carbohydrates into smaller sugar components occurs spontaneously, but the addition of enzymes such as amylases in our saliva makes the reaction occur quickly.

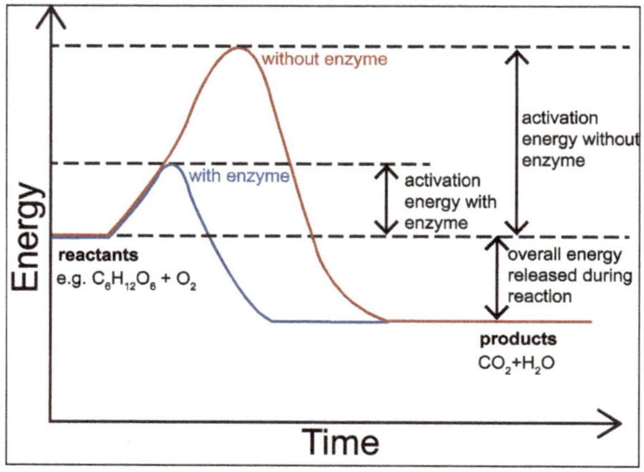

The energy required at each stage of a catalytic reaction.

Enzymes can pair two or more reactions, so that a spontaneous reaction can be used to drive an unfavorable one. For example, the cleavage of the high-energy compound ATP is often used to power other, energetically unfavorable chemical reactions, such as the building of proteins.

Regulation of Enzyme Activity

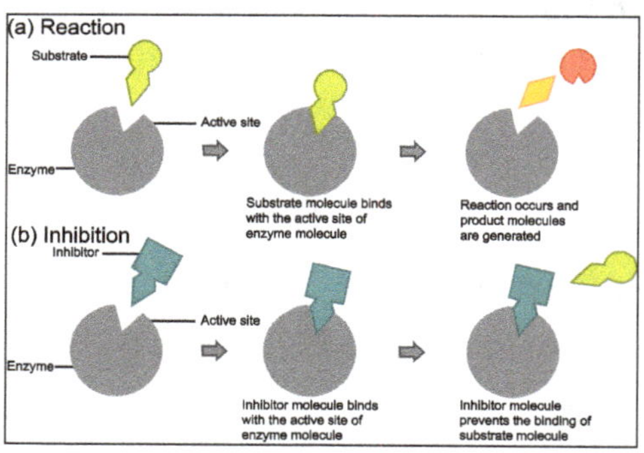

A competitive inhibitor binds reversibly to the active site of the enzyme, preventing the binding of substrate.

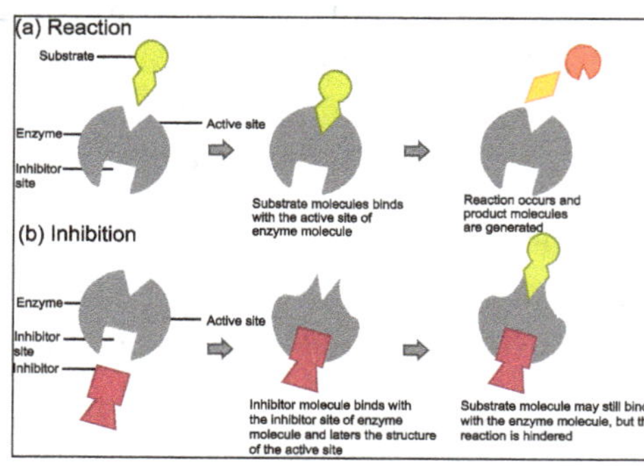

Non-competitive inhibitors do not bind to the active site.

Compounds called inhibitors can decrease enzyme reaction rates through competitive or non-competitive inhibition.

In competitive inhibition, the inhibitor binds directly to the active site, preventing the binding of substrate. The substrate and inhibitor thus "compete" for the active site of the enzyme.

Non-competitive inhibitors do not bind to the active site; rather, they bind to other parts of the enzyme, which can be remote from the active site. The extent of inhibition depends entirely on the inhibitor concentration and will not be affected by the substrate concentration. For example, the poison cyanide combines with the copper prosthetic groups of the enzyme cytochrome c oxidase to inhibit cellular respiration. This type of inhibition is typically irreversible, meaning that the enzyme will no longer function after interacting with the inhibitor.

Some non-competitive inhibitors work by physically blocking the active site. Others bind to the enzyme in a way that alters the three-dimensional structure of the enzyme (its conformation); the change in the enzyme's structure distorts the active site, disabling the enzyme from binding with substrate. In this second form of noncompetitive inhibition, called allosteric inhibition, the inhibitor binds to an allosteric site, changing the shape of the enzyme molecule in a way that prevents it from reacting with the substrate.

Allosteric Control

Allosteric inhibitors are often used to regulate metabolic pathways, in which several enzymes work together in a specific order. In a metabolic pathway, one enzyme takes the product of another enzyme as a substrate. After the catalytic reaction, the product is then passed on to another enzyme. The end product(s) of such a pathway are often allosteric inhibitors for one of the first enzymes of the pathway (usually the first irreversible step, called the committed step), thus regulating the amount of end product made by the pathways. This regulatory process is called negative feedback, because the amount of the end product produced is regulated by its own concentration.

Allosteric molecules can also activate or increase the activity of enzymes by changing the shape of the enzyme's active site in order to facilitate interaction with a substrate. This allosteric control of enzymatic action helps to maintain a stable internal environment in living organisms, by stimulating the production of supplies when needed and preventing the excess manufacture of end products once the demand has been met.

Enzyme-naming Conventions

Enzymes are known for their specificity; that is, they often interact with only one substrate to catalyze a particular reaction. Thus, enzymes have often been named by adding the suffix -ase to the name of the substrate (e.g., lactase is the enzyme that catalyzes the breakdown of lactose). Not all enzymes have been named in this manner, so a more formal method of nomenclature has been developed to classify enzymes.

The International Union of Biochemistry and Molecular Biology has developed a nomenclature for enzymes, called EC numbers. The EC number describes each enzyme using a sequence of four numbers, preceded by "EC." The first number broadly classifies the enzyme based on how it functions to catalyze a reaction.

Under this system, enzymes are broadly organized into six major categories, based on the types of reactions they catalyze:

- EC 1 Oxidoreductases catalyze oxidation/reduction reactions, which involve electron transfer.
- EC 2 Transferases transfer a chemical group called a functional group (e.g., a methyl or phosphate group) from one substance to another.
- EC 3 Hydrolases catalyze the cleavage of chemical bonds through the addition of a water molecule hydrolysis.
- EC 4 Lyases cleave various bonds by means other than hydrolysis and oxidation.
- EC 5 Isomerases transfer a group within a single molecule to form an isomer.
- EC 6 Ligases join two molecules with covalent bonds.

COENZYME

Coenzyme is any of a diverse group of small organic, non-protein, freely diffusing molecules that are loosely associated with and essential for the activity of enzymes, serving as carrier molecules that transfer chemical groups. Coenzymes are sometimes referred to as cosubstrates. These molecules are substrates for enzymes and do not form a permanent part of the enzymes' structures.

The term coenzymes is sometimes defined in such a way as to include prosthetic groups. However, prosthetic groups are non-protein components that are bound tightly (covalently linked) to enzymes—such as iron-sulfur centers, flavin, or haem groups. The International Union of Pure and Applied Chemistry (IUPAC) draws a distinction between coenzymes and prosthetic groups. IUPAC defines a coenzyme as a low-molecular-weight, non-protein organic compound that is loosely attached, participating in enzymatic reactions as a dissociable acceptor of chemical groups or electrons; a prosthetic group is defined as a tightly bound, nonpolypeptide unit in a protein. Both coenzymes and prosthetic groups are types of the broader group of cofactors, which are any non-protein molecules (usually organic molecules or metal ions) that are required by an enzyme for its activity.

Well-known coenzymes include adenosine triphosphate (ATP), which transfers phosphate groups; nicotinamide adenine dinucleotide (NADH, NADPH), which transfers hydrogens and electrons; coenzyme A, which transfers acetyl groups; and S-adenosylmethionine, which transfers methyl groups.

In metabolism, coenzymes are involved in both group-transfer reactions, for example coenzyme A and ATP, and redox reactions, such as coenzyme Q_{10} and NAD^+. Coenzymes are consumed and recycled continuously in metabolism, with one set of enzymes adding a chemical group to the coenzyme and another set removing it. For example, enzymes such as ATP synthase continuously phosphorylate adenosine diphosphate (ADP), converting it into ATP, while enzymes such as kinases dephosphorylate the ATP and convert it back to ADP.

Coenzymes molecules are often vitamins or are made from vitamins. Many coenzymes contain the nucleotide adenosine as part of their structures, such as ATP, coenzyme A, and NAD^+.

Coenzymes are vastly important in life. Some, such as ATP and NADH, form a core part of metablolism and reflect the unity in nature, being present in all known forms of life.

Coenzymes as Metabolic Intermediates

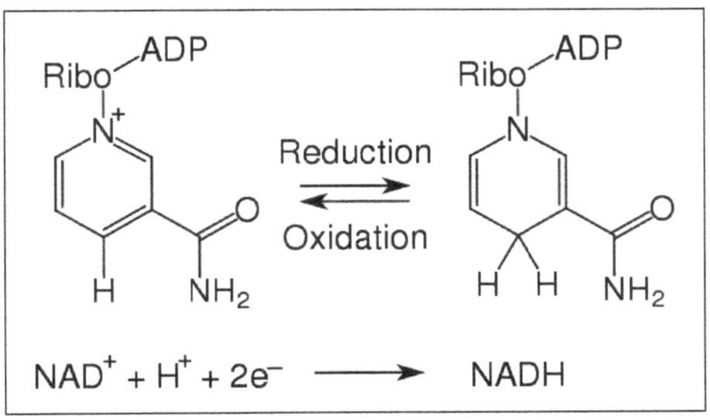

The redox reactions of nicotinamide adenine dinucleotide.

Metabolism involves a vast array of chemical reactions, but most fall under a few basic types of reactions that involve the transfer of functional groups. This common chemistry allows cells to use a small set of metabolic intermediates to carry chemical groups between different reactions. These group-transfer intermediates are the coenzymes.

Each class of group-transfer reaction is carried out by a particular coenzyme, which is the substrate for a set of enzymes that produce it, and a set of enzymes that consume it. An example of this are the dehydrogenases that use nicotinamide adenine dinucleotide (NADH) as a cofactor. Here, hundreds of separate types of enzymes remove electrons from their substrates and reduce NAD+ to NADH. This reduced coenzyme is then a substrate for any of the reductases in the cell that need to reduce their substrates.

Coenzymes are therefore continuously recycled as part of metabolism. As an example, the total quantity of ATP in the human body is about 0.1 mole. This ATP is constantly being broken down into ADP, and then converted back into ATP. Thus, at any given time, the total amount of ATP + ADP remains fairly constant. The energy used by human cells requires the hydrolysis of 100 to 150 moles of ATP daily, which is around 50 to 75 kilograms. Typically, a human will use up their body weight of ATP over the course of the day. This means that each ATP molecule is recycled 1000 to 1500 times daily.

Types

Acting as coenzymes in organisms is the major role of vitamins, although vitamins do have other functions in the body. Coenzymes are also commonly made from nucleotides, such as adenosine triphosphate, the biochemical carrier of phosphate groups, or coenzyme A, the coenzyme that carries acyl groups. Most coenzymes are found in a huge variety of species, and some are universal to all forms of life. An exception to this wide distribution is a group of unique coenzymes that evolved in methanogens, which are restricted to this group of archaea.

Enzyme Structure and Substrate Binding

Amino acid-based enzymes are globular proteins that range in size from less than 100 to more than 2 000 amino acid residues. These amino acids can be arranged as one or more polypeptide chains that are folded and bent to form a specific three-dimensional structure, incorporating a small area known as the active site, where the substrate actually binds. The active site may well involve only a small number (less than 10) of the constituent amino acids.

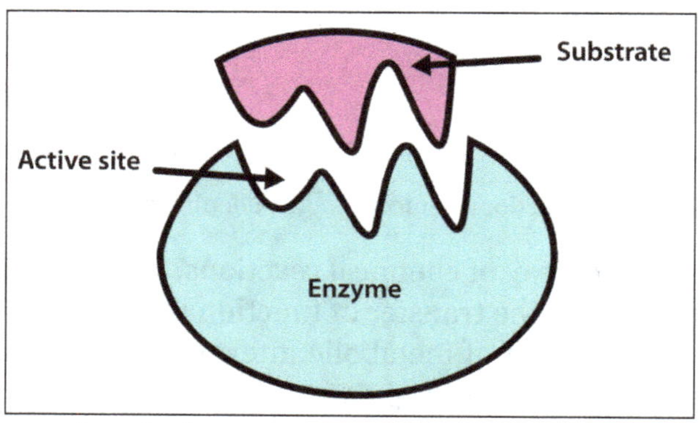

Figure: Representation of substrate binding to the active site of an enzyme molecule.

It is the shape and charge properties of the active site that enable it to bind to a single type of substrate molecule, so that the enzyme is able to demonstrate considerable specificity in its catalytic activity.

The hypothesis that enzyme specificity results from the complementary nature of the substrate and its active site was first proposed by the German chemist Emil Fischer in 1894, and became known as Fischer's 'lock and key hypothesis', whereby only a key of the correct size and shape (the substrate) fits into the keyhole (the active site) of the lock (the enzyme). It is astounding that this theory was proposed at a time when it was not even established that enzymes were proteins. As more was learned about enzyme structure through techniques such as X-ray crystallography, it became clear that enzymes are not rigid structures, but are in fact quite flexible in shape. In the light of this finding, in 1958 Daniel Koshland extended Fischer's ideas and presented the 'induced-fit model' of substrate and enzyme binding, in which the enzyme molecule changes its shape slightly to accommodate the binding of the substrate. The analogy that is commonly used is the 'hand-in-glove model', where the hand and glove are broadly complementary in shape, but the glove is moulded around the hand as it is inserted in order to provide a perfect match.

Since it is the active site alone that binds to the substrate, it is logical to ask what is the role of the rest of the protein molecule. The simple answer is that it acts to stabilize the active site and provide an appropriate environment for interaction of the site with the substrate molecule. Therefore the active site cannot be separated out from the rest of the protein without loss of catalytic activity, although laboratory-based directed (or forced) evolution studies have shown that it is sometimes possible to generate smaller enzymes that do retain activity.

It should be noted that although a large number of enzymes consist solely of protein, many also contain a non-protein component, known as a cofactor, that is necessary for the enzyme's catalytic activity. A cofactor may be another organic molecule, in which case it is called a coenzyme, or it may be an inorganic molecule, typically a metal ion such as iron, manganese, cobalt, copper or zinc. A coenzyme that binds tightly and permanently to the protein is generally referred to as the prosthetic group of the enzyme.

When an enzyme requires a cofactor for its activity, the inactive protein component is generally referred to as an apoenzyme, and the apoenzyme plus the cofactor (i.e. the active enzyme) is called a holoenzyme.

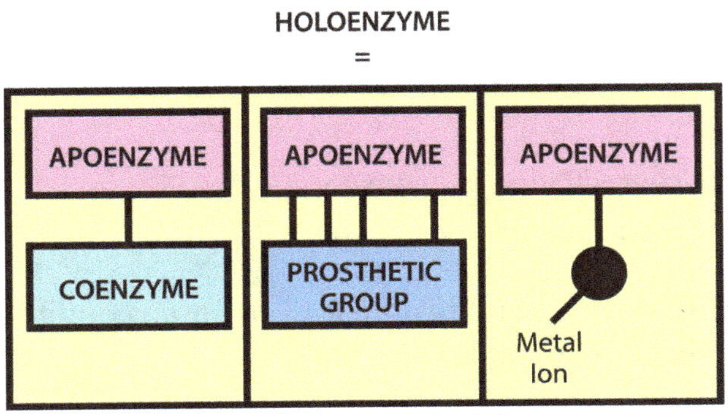

Figure: The components of a holoenzyme.

The need for minerals and vitamins in the human diet is partly attributable to their roles within metabolism as cofactors and coenzymes.

Enzymes and Reaction Equilibrium

How do enzymes work? The broad answer to this question is that they do not alter the equilibrium (i.e. the thermodynamics) of a reaction. This is because enzymes do not fundamentally change the structure and energetics of the products and reagents, but rather they simply allow the reaction equilibrium to be attained more rapidly. Let us therefore begin by clarifying the concept of chemical equilibrium.

In many cases the equilibrium of a reaction is far 'to the right'—that is, virtually all of the substrate (S) is converted into product (P). For this reason, reactions are often written as follows:

$$S \rightarrow P$$

This is a simplification, as in all cases it is more correct to write this reaction as follows:

$$S \rightleftharpoons P$$

This indicates the presence of an equilibrium. To understand this concept it is perhaps most helpful to look at a reaction where the equilibrium point is quite central.

For example:

$$\text{Glucose} \xrightleftharpoons{\text{Glucose isomerase}} \text{Fructose}$$

In this reaction, if we start with a solution of 1 mol l^{-1} glucose and add the enzyme, then upon completion we will have a mixture of approximately 0.5 mol l^{-1} glucose and 0.5 mol l^{-1} fructose. This is the equilibrium point of this particular reaction, and although it may only take a couple of seconds to reach this end point with the enzyme present, we would in fact come to the same point if we put glucose into solution and waited many months for the reaction to occur in the absence of the enzyme. Interestingly, we could also have started this reaction with a 1 mol l^{-1} fructose solution, and it would have proceeded in the opposite direction until the same equilibrium point had been reached.

The equilibrium point for this reaction is expressed by the equilibrium constant K_{eq} as follows:

$$K_{eq} = \frac{\text{Substrate concentration at end point}}{\text{Substrate concentration at end point}} = \frac{0.5}{0.5} = 1$$

Thus for a reaction with central equilibrium, $K_{eq} = 1$, for an equilibrium 'to the right' K_{eq} is >1, and for an equilibrium 'to the left' K_{eq} is <1.

Therefore if a reaction has a K_{eq} value of 10^6, the equilibrium is very far to the right and can be simplified by denoting it as a single arrow. We may often describe this type of reaction as 'going to completion'. Conversely, if a reaction has a K_{eq} value of 10^{-6}, the equilibrium is very far to the left, and for all practical purposes it would not really be considered to proceed at all.

It should be noted that although the concentration of reactants has no effect on the equilibrium point, environmental factors such as pH and temperature can and do affect the position of the equilibrium.

It should also be noted that any biochemical reaction which occurs in vivo in a living system does not occur in isolation, but as part of a metabolic pathway, which makes it more difficult to conceptualize the relationship between reactants and reactions. In vivo reactions are not allowed to proceed to their equilibrium position. If they did, the reaction would essentially stop (i.e. the forward and reverse reactions would balance each other), and there would be no net flux through the pathway. However, in many complex biochemical pathways some of the individual reaction steps are close to equilibrium, whereas others are far from equilibrium, the latter (catalysed by regulatory enzymes) having the greatest capacity to control the overall flux of materials through the pathway.

Enzymes Form Complexes with their Substrates

We often describe an enzyme-catalysed reaction as proceeding through three stages as follows:

$$E + S \rightarrow ES \text{ complex} \rightarrow E + P$$

The ES complex represents a position where the substrate (S) is bound to the enzyme (E) such that the reaction (whatever it might be) is made more favourable. As soon as the reaction has occurred, the product molecule (P) dissociates from the enzyme, which is then free to bind to another substrate molecule. At some point during this process the substrate is converted into an intermediate form (often called the transition state) and then into the product.

The exact mechanism whereby the enzyme acts to increase the rate of the reaction differs from one system to another. However, the general principle is that by binding of the substrate to the enzyme, the reaction involving the substrate is made more favourable by lowering the activation energy of the reaction.

In terms of energetics, reactions can be either exergonic (releasing energy) or endergonic (consuming energy). However, even in an exergonic reaction a small amount of energy, termed the activation energy, is needed to give the reaction a 'kick start.' A good analogy is that of a match, the head of which contains a mixture of energy-rich chemicals (phosphorus sesquisulfide and potassium chlorate). When a match burns it releases substantial amounts of light and heat energy (exergonically reacting with O2 in the air). However, and perhaps fortunately, a match will not spontaneously ignite, but rather a small input of energy in the form of heat generated through friction (i.e. striking of the match) is needed to initiate the reaction. Of course once the match has been struck the amount of energy released is considerable, and greatly exceeds the small energy input during the striking process.

Enzymes are considered to lower the activation energy of a system by making it energetically easier for the transition state to form. In the presence of an enzyme catalyst, the formation of the transition state is energetically more favourable (i.e. it requires less energy for the 'kick start'), thereby accelerating the rate at which the reaction will proceed, but not fundamentally changing the energy levels of either the reactant or the product.

Types of Enzymes

2 CHAPTER

Some of the different types of enzymes are lipase, amylase, trypsin and lactase. Lipases are enzymes which are responsible for the catalysis of the hydrolysis of fats. They perform an integral role in digestion as well as the processing of dietary lipids. The topics elaborated in this chapter will help in gaining a better perspective about these diverse types of enzymes.

PROTEOLYTIC ENZYME

Proteolytic enzyme, also called protease, proteinase, or peptidase are a group of enzymes that break the long chainlike molecules of proteins into shorter fragments (peptides) and eventually into their components, amino acids. Proteolytic enzymes are present in bacteria, archaea, certain types of algae, some viruses, and plants; they are most abundant, however, in animals.

There are different types of proteolytic enzymes, which are classified according to sites at which they catalyze the cleavage of proteins. The two major groups are the exopeptidases, which target the terminal ends of proteins, and the endopeptidases, which target sites within proteins. Endopeptidases employ various catalytic mechanisms; within this group are the aspartic endopeptidases, cysteine endopeptidases, glutamic endopeptidases, metalloendopeptidases, serine endopeptidases, and threonine endopeptidases. The term oligopeptidase is reserved for those enzymes that act specifically on peptides.

Among the best-known proteolytic enzymes are those that reside in the digestive tract. In the stomach, protein materials are attacked initially by a gastric endopeptidase known as pepsin. When the protein material is passed to the small intestine, proteins, which are only partially digested in the stomach, are further attacked by proteolytic enzymes secreted by the pancreas. These enzymes are liberated in the small intestine from inactive precursors produced by the acinar cells in the pancreas. The precursors are called trypsinogen, chymotrypsinogen, proelastase, and procarboxypeptidase. Trypsinogen is transformed to an endopeptidase called trypsin by an enzyme (enterokinase) secreted from the walls of the small intestine. Trypsin then activates the precursors of chymotrypsin, elastase, and carboxypeptidase. When the pancreatic enzymes become activated in the intestine, they convert proteins into free amino acids, which are easily absorbed by the cells of the intestinal wall. The pancreas also produces a protein that inhibits trypsin. It is thought that in this manner the pancreas protects itself from autodigestion.

LIPASE

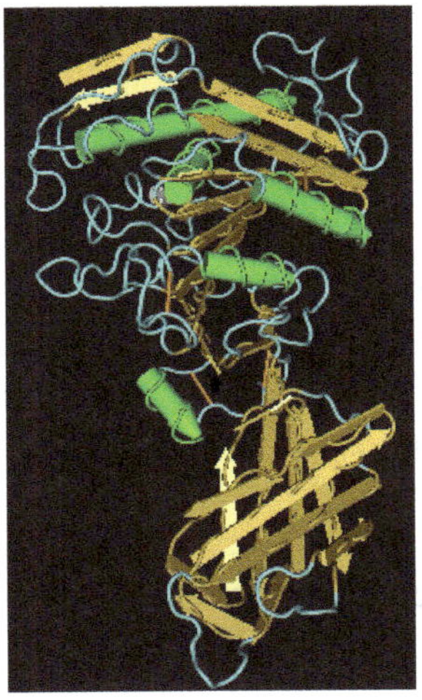

A computer-generated image of a type of pancreatic lipase (PLRP2) from the guinea pig.

A lipase is any enzyme that catalyzes the hydrolysis of fats (lipids). Lipases are a subclass of the esterases.

Lipases perform essential roles in digestion, transport and processing of dietary lipids (e.g. triglycerides, fats, oils) in most, if not all, living organisms. Genes encoding lipases are even present in certain viruses.

Most lipases act at a specific position on the glycerol backbone of a lipid substrate (A1, A2 or A3)(small intestine). For example, human pancreatic lipase (HPL), which is the main enzyme that breaks down dietary fats in the human digestive system, converts triglyceride substrates found in ingested oils to monoglycerides and two fatty acids.

Several other types of lipase activities exist in nature, such as phospholipases and sphingomyelinases; however, these are usually treated separately from "conventional" lipases.

Some lipases are expressed and secreted by pathogenic organisms during an infection. In particular, *Candida albicans* has many different lipases, possibly reflecting broad-lipolytic activity, which may contribute to the persistence and virulence of *C. albicans* in human tissue.

Structure and Catalytic Mechanism

A diverse array of genetically distinct lipase enzymes are found in nature, and they represent several types of protein folds and catalytic mechanisms. However, most are built on an alpha/beta hydrolase fold and employ a chymotrypsin-like hydrolysis mechanism using a catalytic triad consisting of a serine nucleophile, a histidine base, and an acid residue, usually aspartic acid.

Physiological distribution

Lipases are involved in diverse biological processes which range from routine metabolism of dietary triglycerides to cell signaling and inflammation. Thus, some lipase activities are confined to specific compartments within cells while others work in extracellular spaces:

- In the example of lysosomal lipase, the enzyme is confined within an organelle called the lysosome.
- Other lipase enzymes, such as pancreatic lipases, are secreted into extracellular spaces where they serve to process dietary lipids into more simple forms that can be more easily absorbed and transported throughout the body.
- Fungi and bacteria may secrete lipases to facilitate nutrient absorption from the external medium (or in examples of pathogenic microbes, to promote invasion of a new host).
- Certain wasp and bee venoms contain phospholipases that enhance the effects of injury and inflammation delivered by a sting.
- As biological membranes are integral to living cells and are largely composed of phospholipids, lipases play important roles in cell biology.
- *Malassezia globosa*, a fungus thought to be the cause of human dandruff, uses lipase to break down sebum into oleic acid and increase skin cell production, causing dandruff.

Human Lipases

The main lipases of the human digestive system are pancreatic lipase (PL) and pancreatic lipase related protein 2 (PLRP2), which are secreted by the pancreas. Humans also have several related enzymes, including hepatic lipase, endothelial lipase, and lipoprotein lipase. Not all of these lipases function in the gut.

Name	Gene	Location	Description
Bile salt-dependent lipase	bsdl	Pancreas, breast milk	Aids in the digestion of fats
Pancreatic lipase	*PNLIP*	Digestive juice	In order to exhibit optimal enzyme activity in the gut lumen, PL requires another protein, colipase, which is also secreted by the pancreas.

Lysosomal lipase	*LIPA*	Interior space of organelle: lysosome	Also referred to as lysosomal acid lipase (LAL or LIPA) or acid cholesteryl ester hydrolase
Hepatic lipase	*LIPC*	Endothelium	Hepatic lipase acts on the remaining lipids carried on lipoproteins in the blood to regenerate LDL (low density lipoprotein).
Lipoprotein lipase	*LPL* or "LIPD"	Endothelium	Lipoprotein lipase functions in the blood to act on triacylglycerides carried on VLDL (very low density lipoprotein) so that cells can take up the freed fatty acids.
Gastric lipase	*LIPF*	Digestive juice	Functions in the infant at a near-neutral pH to aid in the digestion of lipids
Pancreatic lipase related protein 1	*PNLIPRP1* or "PLRP1"	Digestive juice	Pancreatic lipase related protein 1 is very similar to PLRP2 and PL by amino acid sequence (all three genes probably arose via gene duplication of a single ancestral pancreatic lipase gene). However, PLRP1 is devoid of detectable lipase activity and its function remains unknown, even though it is conserved in other mammals.
Lingual lipase	?	Saliva	Active at gastric pH levels. Optimum pH is about 3.5-6. Secreted by several of the salivary glands (Ebner's glands at the back of the tongue (lingua), the sublingual glands, and the parotid glands)

Other lipases include *LIPH, LIPI, LIPJ, LIPK, LIPM, LIPN, MGLL, DAGLA, DAGLB,* and *CEL.*

There also are a diverse array of phospholipases, but these are not always classified with the other lipases.

Industrial uses

Lipases serve important roles in human practices as ancient as yogurt and cheese fermentation. However, lipases are also being exploited as cheap and versatile catalysts to degrade lipids in more modern applications. For instance, a biotechnology company has brought recombinant lipase enzymes to market for use in applications such as baking, laundry detergents and even as biocatalysts in alternative energy strategies to convert vegetable oil into fuel. High enzyme activity lipase can replace traditional catalyst in processing biodiesel, as this enzyme replaces chemicals in a process which is otherwise highly energy intensive, and can be more environmentally friendly and safe. Industrial application of lipases requires process intensification for continuous processing using tools like continuous flow microreactors at small scale. Lipases are generally animal sourced, but can also be sourced microbially.

Diagnostic use

Blood tests for lipase may be used to help investigate and diagnose acute pancreatitis and other disorders of the pancreas. Measured serum lipase values may vary depending on the method of analysis.

Medical use

Lipase can also assist in the breakdown of fats into lipids in those undergoing pancreatic enzyme replacement therapy (PERT). It is a key component in Sollpura (Liprotamase).

AMYLASE

Amylase is an enzyme that catalyses the hydrolysis of starch into sugars. Amylase is present in the saliva of humans and some other mammals, where it begins the chemical process of digestion. Foods that contain large amounts of starch but little sugar, such as rice and potatoes, may acquire a slightly sweet taste as they are chewed because amylase degrades some of their starch into sugar. The pancreas and salivary gland make amylase (alpha amylase) to hydrolyse dietary starch into disaccharides and trisaccharides which are converted by other enzymes to glucose to supply the body with energy. Plants and some bacteria also produce amylase. As *diastase*, amylase was the first enzyme to be discovered and isolated. All amylases are glycoside hydrolases and act on α-1,4-glycosidic bonds.

Classification

	α-amylase	β-amylase	γ-amylase
Source	Animals, plants, microbes	Plants, microbes	Animals, microbes
Tissue	Saliva, pancreas	Seeds, fruits	Small intestine
Cleavage site	Random α-1,4 glycosidic bond	Second α-1,4 glycosidic bond	Last α-1,4 glycosidic bond
Reaction products	Maltose, dextrin, etc	Maltose	Glucose
Optimum pH	6.7–7.0	4.0–5.0	3.0
Optimum temperature in brewing	63–70 °C	55–65 °C	

α-Amylase

The α-amylases (alternative names: 1,4-α-D-glucan glucanohydrolase; glycogenase) are calcium metalloenzymes. By acting at random locations along the starch chain, α-amylase breaks down long-chain saccharides, ultimately yielding either maltotriose and maltose from amylose, or maltose, glucose and "limit dextrin" from amylopectin.

Because it can act anywhere on the substrate, α-amylase tends to be faster-acting than β-amylase. In animals, it is a major digestive enzyme, and its optimum pH is 6.7–7.0.

In human physiology, both the salivary and pancreatic amylases are α-amylases.

The α-amylase form is also found in plants, fungi (ascomycetes and basidiomycetes) and bacteria (*Bacillus*).

β-Amylase

Another form of amylase, β-amylase (alternative names: 1,4-α-D-glucan maltohydrolase; glycogenase; saccharogen amylase) is also synthesized by bacteria, fungi, and plants. Working from the non-reducing end, β-amylase catalyzes the hydrolysis of the second α-1,4 glycosidic bond, cleaving off two glucose units (maltose) at a time. During the ripening of fruit, β-amylase breaks starch into maltose, resulting in the sweet flavor of ripe fruit.

Both α-amylase and β-amylase are present in seeds; β-amylase is present in an inactive form prior to germination, whereas α-amylase and proteases appear once germination has begun. Many microbes also produce amylase to degrade extracellular starches. Animal tissues do not contain β-amylase, although it may be present in microorganisms contained within the digestive tract. The optimum pH for β-amylase is 4.0–5.0

γ-Amylase

γ-Amylase (alternative names: glucan 1,4-α-glucosidase; amyloglucosidase; *exo*-1, 4-α-glucosidase; glucoamylase; lysosomal α-glucosidase; 1,4-α-D-glucan glucohydrolase) will cleave α(1–6) glycosidic linkages, as well as the last α-1,4 glycosidic bond at the nonreducing end of amylose and amylopectin, yielding glucose. The γ-amylase has most acidic optimum pH of all amylases because it is most active around pH 3.

Uses

Fermentation

α- and β-amylases are important in brewing beer and liquor made from sugars derived from starch. In fermentation, yeast ingests sugars and excretes ethanol. In beer and some liquors, the sugars present at the beginning of fermentation have been produced by "mashing" grains or other starch sources (such as potatoes). In traditional beer brewing, malted barley is mixed with hot water to create a "mash", which is held at a given temperature to allow the amylases in the malted grain to convert the barley's starch into sugars. Different temperatures optimize the activity of alpha or beta amylase, resulting in different mixtures of fermentable and unfermentable sugars. In selecting mash temperature and grain-to-water ratio, a brewer can change the alcohol content, mouthfeel, aroma, and flavor of the finished beer.

In some historic methods of producing alcoholic beverages, the conversion of starch to sugar starts with the brewer chewing grain to mix it with saliva. This practice is no longer widely in use.

Flour Additive

Amylases are used in breadmaking and to break down complex sugars, such as starch (found in flour), into simple sugars. Yeast then feeds on these simple sugars and converts it into the waste products of ethanol and carbon dioxide. This imparts flavour and causes the bread to rise. While amylases are found naturally in yeast cells, it takes time for the yeast to produce enough of these enzymes to break down significant quantities of starch in the bread. This is the reason for long fermented doughs such as sourdough. Modern breadmaking techniques have included amylases (often in the form of malted barley) into bread improver, thereby making the process faster and more practical for commercial use.

α-Amylase is often listed as an ingredient on commercially package-milled flour. Bakers with long exposure to amylase-enriched flour are at risk of developing dermatitis or asthma.

Molecular Biology

In molecular biology, the presence of amylase can serve as an additional method of selecting for successful integration of a reporter construct in addition to antibiotic resistance. As reporter genes are flanked by homologous regions of the structural gene for amylase, successful integration will disrupt the amylase gene and prevent starch degradation, which is easily detectable through iodine staining.

Medical uses

Amylase also has medical applications in the use of pancreatic enzyme replacement therapy (PERT). It is one of the components in Sollpura (liprotamase) to help in the breakdown of saccharides into simple sugars.

Other uses

An inhibitor of alpha-amylase, called phaseolamin, has been tested as a potential diet aid. When used as a food additive, amylase has E number E1100, and may be derived from pig pancreas or mold fungi. Bacilliary amylase is also used in clothing and dishwasher detergents to dissolve starches from fabrics and dishes.

Factory workers who work with amylase for any of the above uses are at increased risk of occupational asthma. Five to nine percent of bakers have a positive skin test, and a fourth to a third of bakers with breathing problems are hypersensitive to amylase.

Hyperamylasemia

Blood serum amylase may be measured for purposes of medical diagnosis. A higher than normal concentration may reflect one of several medical conditions, including acute inflammation of the pancreas (it may be measured concurrently with the more specific lipase), but also perforated peptic ulcer, torsion of an ovarian cyst, strangulation, ileus, mesenteric ischemia, macroamylasemia and mumps. Amylase may be measured in other body fluids, including urine and peritoneal fluid.

A January 2007 study from Washington University in St. Louis suggests that saliva tests of the enzyme could be used to indicate sleep deficits, as the enzyme increases its activity in correlation with the length of time a subject has been deprived of sleep.

TRYPSIN

Trypsin is a serine protease from the PA clan superfamily, found in the digestive system of many vertebrates, where it hydrolyzes proteins. Trypsin is formed in the small intestine when its proenzyme form, the trypsinogen produced by the pancreas, is activated. Trypsin cleaves peptide chains mainly at the carboxyl side of the amino acids lysine or arginine. It is used for numerous biotechnological processes. The process is commonly referred to as trypsin proteolysis or trypsinisation, and proteins that have been digested/treated with trypsin are said to have been trypsinized. Trypsin was discovered in 1876 by Wilhelm Kühne.

Function

In the duodenum, trypsin catalyzes the hydrolysis of peptide bonds, breaking down proteins into smaller peptides. The peptide products are then further hydrolyzed into amino acids via other proteases, rendering them available for absorption into the blood stream. Tryptic digestion is a necessary step in protein absorption, as proteins are generally too large to be absorbed through the lining of the small intestine.

Trypsin is produced as the inactive zymogen trypsinogen in the pancreas. When the pancreas is stimulated by cholecystokinin, it is then secreted into the first part of the small intestine (the duodenum) via the pancreatic duct. Once in the small intestine, the enzyme enteropeptidase activates trypsinogen into trypsin by proteolytic cleavage. Autocatalysis does not happen with trypsin, as trypsinogen is a poor substrate, therefore enzymatic damage to the pancreas is avoided.

Mechanism

The enzymatic mechanism is similar to that of other serine proteases. These enzymes contain a catalytic triad consisting of histidine-57, aspartate-102, and serine-195. This catalytic triad was formerly called a charge relay system, implying the abstraction of protons from serine to histidine and from histidine to aspartate, but owing to evidence provided by NMR that the resultant alkoxide form of serine would have a much stronger pull on the proton than does the imidazole ring of histidine, current thinking holds instead that serine and histidine each have effectively equal share of the proton, forming short low-barrier hydrogen bonds therewith. By these means, the nucleophilicity of the active site serine is increased, facilitating its attack on the amide carbon during proteolysis. The enzymatic reaction that trypsin catalyzes is thermodynamically favorable, but requires significant activation energy (it is "kinetically unfavorable"). In addition, trypsin contains an "oxyanion hole" formed by the backbone amide hydrogen atoms of Gly-193 and Ser-195, which through hydrogen bonding stabilize the negative charge which accumulates on the amide oxygen after nucleophilic attack on the planar amide carbon by the serine oxygen causes that carbon to assume a tetrahedral geometry. Such stabilisation of this tetrahedral intermediate helps to reduce the energy barrier of its formation and is concomitant with a lowering of the free energy of the transition state. Preferential binding of the transition state is a key feature of enzyme chemistry.

The aspartate residue (Asp 189) located in the catalytic pocket (S1) of trypsin is responsible for attracting and stabilizing positively charged lysine and/or arginine, and is, thus, responsible for the specificity of the enzyme. This means that trypsin predominantly cleaves proteins at the carboxyl side (or "C-terminal side") of the amino acids lysine and arginine except when either is bound to a C-terminal proline, although large-scale mass spectrometry data suggest cleavage occurs even with proline. Trypsin is considered an endopeptidase, i.e., the cleavage occurs within the polypeptide chain rather than at the terminal amino acids located at the ends of polypeptides.

Properties

Human trypsin has an optimal operating temperature of about 37 °C. In contrast, the Atlantic cod has several types of trypsins for the poikilotherm fish to survive at different body temperatures. Cod trypsins include trypsin I with an activity range of 4 to 65 °C (40 to 150 °F) and maximal activity at 55 °C (130 °F), as well as trypsin Y with a range of 2 to 30 °C (36 to 86 °F) and a maximal activity at 21 °C (70 °F).

As a protein, trypsin has various molecular weights depending on the source. For example, a molecular weight of 23.3 kDa is reported for trypsin from bovine and porcine sources.

The activity of trypsin is not affected by the enzyme inhibitor tosyl phenylalanyl chloromethyl ketone, TPCK, which deactivates chymotrypsin. This is important because, in some applications, like mass spectrometry, the specificity of cleavage is important.

Trypsin should be stored at very cold temperatures (between −20 and −80 °C) to prevent autolysis, which may also be impeded by storage of trypsin at pH 3 or by using trypsin modified by reductive methylation. When the pH is adjusted back to pH 8, activity returns.

Isozymes

These human genes encode proteins with trypsin enzymatic activity:

protease, serine, 1 (trypsin 1)	
Identifiers	
Symbol	PRSS1
Alt. symbols	TRY1
NCBI gene	5644
HGNC	9475
OMIM	276000
RefSeq	NM_002769
UniProt	P07477
Other data	
EC number	3.4.21.4
Locus	Chr. 7 *q32-qter*

protease, serine, 2 (trypsin 2)	
Identifiers	
Symbol	PRSS2
Alt. symbols	TRYP2
NCBI gene	5645
HGNC	9483
OMIM	601564
RefSeq	NM_002770
UniProt	P07478
Other data	
EC number	3.4.21.4
Locus	Chr. 7 *q35*

protease, serine, 3 (mesotrypsin)	
Identifiers	
Symbol	PRSS3
Alt. symbols	PRSS4
NCBI gene	5646
HGNC	9486
OMIM	613578
RefSeq	NM_002771
UniProt	P35030
Other data	
EC number	3.4.21.4
Locus	Chr. 9 *p13*

Other isoforms of trypsin may also be found in other organisms.

Clinical Significance

Activation of trypsin from proteolytic cleavage of trypsinogen in the pancreas can lead to a series of events that cause pancreatic self-digestion, resulting in pancreatitis. One consequence of the autosomal recessive disease cystic fibrosis is a deficiency in transport of trypsin and other digestive enzymes from the pancreas. This leads to the disorder termed meconium ileus, which involves intestinal obstruction (ileus) due to overly thick meconium, which is normally broken down by trypsin and other proteases, then passed in feces.

Applications

Trypsin is available in high quantity in pancreases, and can be purified rather easily. Hence, it has been used widely in various biotechnological processes.

In a tissue culture lab, trypsin is used to resuspend cells adherent to the cell culture dish wall during the process of harvesting cells. Some cell types adhere to the sides and bottom of a dish when cultivated *in vitro*. Trypsin is used to cleave proteins holding the cultured cells to the dish, so that the cells can be removed from the plates.

Trypsin can also be used to dissociate dissected cells (for example, prior to cell fixing and sorting).

Trypsin can be used to break down casein in breast milk. If trypsin is added to a solution of milk powder, the breakdown of casein causes the milk to become translucent. The rate of reaction can be measured by using the amount of time needed for the milk to turn translucent.

Trypsin is commonly used in biological research during proteomics experiments to digest proteins into peptides for mass spectrometry analysis, e.g. in-gel digestion. Trypsin is particularly suited for this, since it has a very well defined specificity, as it hydrolyzes only the peptide bonds in which the carbonyl group is contributed either by an arginine or lysine residue.

Trypsin can also be used to dissolve blood clots in its microbial form and treat inflammation in its pancreatic form.

In Food

Commercial protease preparations usually consist of a mixture of various protease enzymes that often includes trypsin. These preparations are widely used in food processing:

- As a baking enzyme to improve the workability of dough.

- In the extraction of seasonings and flavorings from vegetable or animal proteins and in the manufacture of sauces.
- To control aroma formation in cheese and milk products.
- To improve the texture of fish products.
- To tenderize meat.
- During cold stabilization of beer.
- In the production of hypoallergenic food where proteases break down specific allergenic proteins into nonallergenic peptides, for example, proteases are used to produce hypoallergenic baby food from cow's milk, thereby diminishing the risk of babies developing milk allergies.

Trypsin Inhibitor

To prevent the action of active trypsin in the pancreas, which can be highly damaging, inhibitors such as BPTI and SPINK1 in the pancreas and α1-antitrypsin in the serum are present as part of the defense against its inappropriate activation. Any trypsin prematurely formed from the inactive trypsinogen is then bound by the inhibitor. The protein-protein interaction between trypsin and its inhibitors is one of the tightest bound, and trypsin is bound by some of its pancreatic inhibitors nearly irreversibly. In contrast with nearly all known protein assemblies, some complexes of trypsin bound by its inhibitors do not readily dissociate after treatment with 8M urea.

LACTASE

Lactase is an enzyme produced by many organisms. It is located in the brush border of the small intestine of humans and other mammals. Lactase is essential to the complete digestion of whole milk; it breaks down lactose, a sugar which gives milk its sweetness. Lacking lactase, a person consuming dairy products may experience the symptoms of lactose intolerance. Lactase can be purchased as a food supplement, and is added to milk to produce "lactose-free" milk products.

Lactase (also known as lactase-phlorizin hydrolase, or LPH), a part of the β-galactosidase family of enzymes, is a glycoside hydrolase involved in the hydrolysis of the disaccharide lactose into constituent galactose and glucose monomers. Lactase is present predominantly along the brush border membrane of the differentiated enterocytes lining the villi of the small intestine. In humans, lactase is encoded by the LCT gene.

Uses

Food use

Lactase is an enzyme that some people are unable to produce in their small intestine. Without it they can't break down the natural lactose in milk, leaving them with diarrhea, gas and bloating when drinking regular milk. Technology to produce lactose-free milk, ice cream and yogurt was developed by the USDA Agricultural Research Service in 1985. This technology is used to add lactase to milk, thereby hydrolyzing the lactose naturally found in milk, leaving it slightly sweet but digestible by everyone. Without lactase, lactose intolerant people pass the lactose undigested to the colon where bacteria break it down, creating carbon dioxide and that leads to bloating and flatulence.

Medical use

Lactase supplements are sometimes used to treat lactose intolerance.

Industrial use

Lactase produced commercially can be extracted both from yeasts such as *Kluyveromyces fragilis* and *Kluyveromyces lactis* and from molds, such as *Aspergillus niger* and *Aspergillus oryzae*. Its primary commercial use, in supplements such as Lacteeze and Lactaid, is to break down lactose in milk to make it suitable for people with lactose intolerance, However, the U.S. Food and Drug Administration has not formally evaluated the effectiveness of these products.

Lactase is also used to screen for blue white colonies in the multiple cloning sites of various plasmid vectors in *Escherichia coli* or other bacteria.

Mechanism

The optimum temperature for human lactase is about 37 °C for its activity and the optimum pH is 6.

In metabolism, the β-glycosidic bond in D-lactose is hydrolyzed to form D-galactose and D-glucose, which can be absorbed through the intestinal walls and into the bloodstream. The overall reaction that lactase catalyzes is $C_{12}H_{22}O_{11} + H_2O \rightarrow C_6H_{12}O_6 + C_6H_{12}O_6$ + heat.

The catalytic mechanism of D-lactose hydrolysis retains the substrate anomeric configuration in the products. While the details of the mechanism are uncertain, the stereochemical retention is achieved through a double displacement reaction. Studies of *E. coli* lactase have proposed that hydrolysis is initiated when a glutamate nucleophile on the enzyme attacks from the axial side of the galactosyl carbon in the β-glycosidic bond. The removal of the D-glucose leaving group may be facilitated by Mg-dependent acid catalysis. The enzyme is liberated from the α-galactosyl moiety upon equatorial nucleophilic attack by water, which produces D-galactose.

Substrate modification studies have demonstrated that the 3′-OH and 2′-OH moieties on the galactopyranose ring are essential for enzymatic recognition and hydrolysis. The 3′-hydroxy group is involved in initial binding to the substrate while the 2′- group is not necessary for recognition but needed in subsequent steps. This is demonstrated by the fact that a 2-deoxy analog is an effective competitive inhibitor (K_i = 10mM). Elimination of specific hydroxyl groups on the glucopyranose moiety does not completely eliminate catalysis.

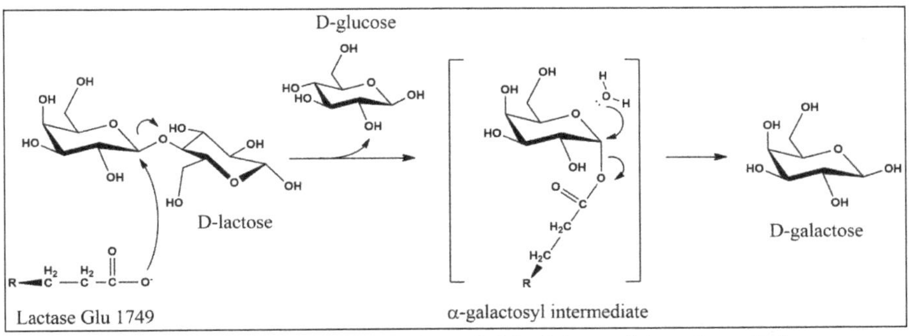

Lactase also catalyzes the conversion of phlorizin to phloretin and glucose.

Structure and Biosynthesis

Preprolactase, the primary translation product, has a single polypeptide primary structure consisting of 1927 amino acids. It can be divided into five domains: (i) a 19-amino-acid cleaved signal sequence; (ii) a large prosequence domain that is not present in mature lactase; (iii) the mature lactase segment; (iv) a membrane-spanning hydrophobic anchor; and (v) a short hydrophilic carboxyl terminus. The signal sequence is cleaved in the endoplasmic reticulum, and the resulting 215-kDa pro-LPH is sent to the Golgi apparatus, where it is heavily glycosylated and proteolytically processed to its mature form. The prodomain has been shown to act as an intramolecular chaperone in the ER, preventing trypsin cleavage and allowing LPH to adopt the necessary 3-D structure to be transported to the Golgi apparatus.

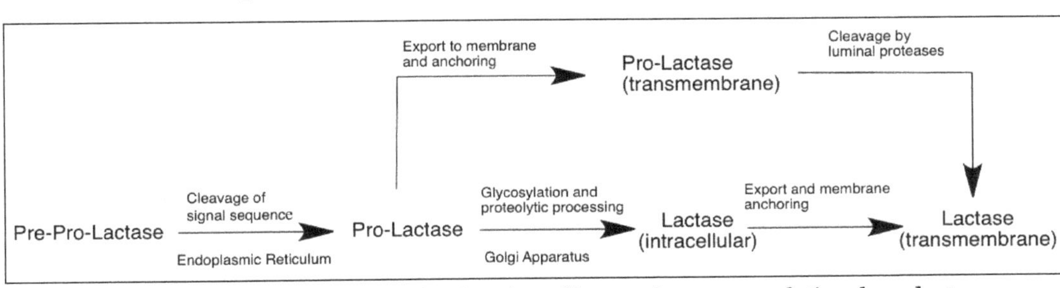

Schematic of processing and localization of human lactase translational product.

Mature human lactase consists of a single 160-kDa polypeptide chain that localizes to the brush border membrane of intestinal epithelial cells. It is oriented with the

N-terminus outside the cell and the C-terminus in the cytosol. LPH contains two catalytic glutamic acid sites. In the human enzyme, the lactase activity has been connected to Glu-1749, while Glu-1273 is the site of phlorizin hydrolase function.

Genetic Expression and Regulation

Lactase is encoded by a single genetic locus on chromosome 2. It is expressed exclusively by mammalian small intestine enterocytes and in very low levels in the colon during fetal development. Humans are born with high levels of lactase expression. In most of the world's population, lactase transcription is down-regulated after weaning, resulting in diminished lactase expression in the small intestine, which causes the common symptoms of adult-type hypolactasia, or lactose intolerance.

Some population segments exhibit lactase persistence resulting from a mutation that is postulated to have occurred 5,000–10,000 years ago, coinciding with the rise of cattle domestication. This mutation has allowed almost half of the world's population to metabolize lactose without symptoms. Studies have linked the occurrence of lactase persistence to two different single-nucleotide polymorphisms about 14 and 22 kilobases upstream of the 5′-end of the LPH gene. Both mutations, C→T at position -13910 and G→ A at position -22018, have been independently linked to lactase persistence.

The lactase promoter is 150 base pairs long and is located just upstream of the site of transcription initiation. The sequence is highly conserved in mammals, suggesting that critical cis-transcriptional regulators are located nearby. Cdx-2, HNF-1α, and GATA have been identified as transcription factors. Studies of hypolactasia onset have demonstrated that despite polymorphisms, little difference exists in lactase expression in infants, showing that the mutations become increasingly relevant during development. Developmentally regulated DNA-binding proteins may down-regulate transcription or destabilize mRNA transcripts, causing decreased LPH expression after weaning.

ACETYLCHOLINESTERASE

Acetylcholinesterase, also known as AChE or acetylhydrolase, is the primary cholinesterase in the body. It is an enzyme that catalyzes the breakdown of acetylcholine and of some other choline esters that function as neurotransmitters. AChE is found at mainly neuromuscular junctions and in chemical synapses of the cholinergic type, where its activity serves to terminate synaptic transmission. It belongs to carboxylesterase family of enzymes. It is the primary target of inhibition by organophosphorus compounds such as nerve agents and pesticides.

Enzyme Structure and Mechanism

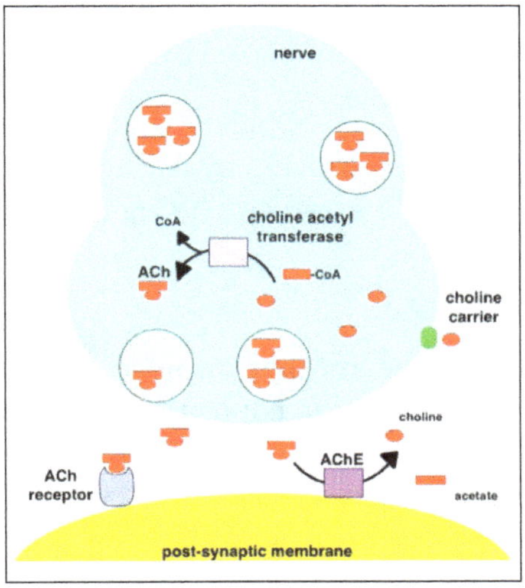

AChe mechanism of action.

AChE is a hydrolase that hydrolyzes choline esters. It has a very high catalytic activity—each molecule of AChE degrades about 25,000 molecules of acetylcholine (ACh) per second, approaching the limit allowed by diffusion of the substrate. The active site of AChE comprises 2 subsites—the anionic site and the esteratic subsite. The structure and mechanism of action of AChE have been elucidated from the crystal structure of the enzyme.

The anionic subsite accommodates the positive quaternary amine of acetylcholine as well as other cationic substrates and inhibitors. The cationic substrates are not bound by a negatively charged amino acid in the anionic site, but by interaction of 14 aromatic residues that line the gorge leading to the active site. All 14 amino acids in the aromatic gorge are highly conserved across different species. Among the aromatic amino acids, tryptophan 84 is critical and its substitution with alanine results in a 3000-fold decrease in reactivity. The gorge penetrates halfway through the enzyme and is approximately 20 angstroms long. The active site is located 4 angstroms from the bottom of the molecule.

The esteratic subsite, where acetylcholine is hydrolyzed to acetate and choline, contains the catalytic triad of three amino acids: serine 200, histidine 440 and glutamate 327. These three amino acids are similar to the triad in other serine proteases except that the glutamate is the third member rather than aspartate. Moreover, the triad is of opposite chirality to that of other proteases. The hydrolysis reaction of the carboxyl ester leads to the formation of an acyl-enzyme and free choline. Then, the acyl-enzyme undergoes nucleophilic attack by a water molecule, assisted by the histidine 440 group, liberating acetic acid and regenerating the free enzyme.

Biological Function

During neurotransmission, ACh is released from the presynaptic neuron into the synaptic cleft and binds to ACh receptors on the post-synaptic membrane, relaying the signal from the nerve. AChE, also located on the post-synaptic membrane, terminates the signal transmission by hydrolyzing ACh. The liberated choline is taken up again by the pre-synaptic neuron and ACh is synthesized by combining with acetyl-CoA through the action of choline acetyltransferase.

A cholinomimetic drug disrupts this process by acting as a cholinergic neurotransmitter that is impervious to acetylcholinesterase's lysing action.

Disease Relevance

For a cholinergic neuron to receive another impulse, ACh must be released from the ACh receptor. This occurs only when the concentration of ACh in the synaptic cleft is very low. Inhibition of AChE leads to accumulation of ACh in the synaptic cleft and results in impeded neurotransmission.

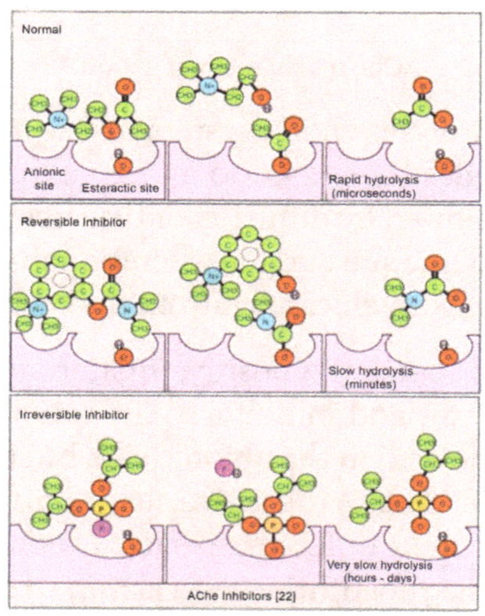

Mechanism of Inhibitors of AChE.

Irreversible inhibitors of AChE may lead to muscular paralysis, convulsions, bronchial constriction, and death by asphyxiation. Organophosphates (OP), esters of phosphoric acid, are a class of irreversible AChE inhibitors. Cleavage of OP by AChE leaves a phosphoryl group in the esteratic site, which is slow to be hydrolyzed (on the order of days) and can become covalently bound. Irreversible AChE inhibitors have been used in insecticides (e.g., malathion) and nerve gases for chemical warfare (e.g., Sarin and Soman). Carbamates, esters of N-methyl carbamic acid, are AChE inhibitors that hydrolyze in hours and have been used for medical purposes (e.g., physostigmine for the treatment of glaucoma). Reversible inhibitors occupy the esteratic site for short periods of time (seconds to minutes) and are used to treat of a range of central nervous system diseases. Tetrahydroaminoacridine (THA) and donepezil are FDA-approved to improve cognitive function in Alzheimer's disease. Rivastigmine is also used to treat Alzheimer's and Lewy body dementia, and pyridostigmine bromide is used to treat myasthenia gravis.

An endogenous inhibitor of AChE in neurons is Mir-132 microRNA, which may limit inflammation in the brain by silencing the expression of this protein and allowing ACh to act in an anti-inflammatory capacity.

It has also been shown that the main active ingredient in cannabis, tetrahydrocannabinol, is a competitive inhibitor of acetylcholinesterase.

Distribution

AChE is found in many types of conducting tissue: nerve and muscle, central and peripheral tissues, motor and sensory fibers, and cholinergic and noncholinergic fibers. The activity of AChE is higher in motor neurons than in sensory neurons.

Acetylcholinesterase is also found on the red blood cell membranes, where different forms constitute the Yt blood group antigens. Acetylcholinesterase exists in multiple molecular forms, which possess similar catalytic properties, but differ in their oligomeric assembly and mode of attachment to the cell surface.

AChE Gene

In mammals, acetylcholinesterase is encoded by a single AChE gene while some invertebrates have multiple acetylcholinesterase genes. Note higher vertebrates also encode a closely related paralog BCHE (butyrylcholinesterase) with 50% amino acid identity to ACHE. Diversity in the transcribed products from the sole mammalian gene arises from alternative mRNA splicing and post-translational associations of catalytic and structural subunits. There are three known forms: T (tail), R (read through), and H(hydrophobic).

$AChE_T$

The major form of acetylcholinesterase found in brain, muscle, and other tissues, known as is the hydrophilic species, which forms disulfide-linked oligomers with collagenous, or lipid-containing structural subunits. In the neuromuscular junctions AChE expresses in asymmetric form which associates with ColQ or subunit. In the central nervous system it is associated with PRiMA which stands for Proline Rich Membrane anchor to form symmetric form. In either case, the ColQ or PRiMA anchor serves to maintain the enzyme in the intercellular junction, ColQ for the neuromuscular junction and PRiMA for synapses.

$AChE_H$

The other, alternatively spliced form expressed primarily in the erythroid tissues, differs at the C-terminus, and contains a cleavable hydrophobic peptide with a PI-anchor site. It associates with membranes through the phosphoinositide (PI) moieties added post-translationally.

AChE$_R$

The third type has, so far, only been found in *Torpedo* sp. and mice although it is hypothesized in other species. It is thought to be involved in the stress response and, possibly, inflammation.

Nomenclature

The nomenclatural variations of ACHE and of cholinesterases generally are discussed at *Cholinesterase Types and nomenclature.*

Inhibitors

For acetylcholine esterase (AChE), reversible inhibitors are those that do not irreversibly bond to and deactivate AChE. Drugs that reversibly inhibit acetylcholine esterase are being explored as treatments for Alzheimer's disease and myasthenia gravis, among others. Examples include tacrine and donepezil.

HELICASE

Helicases are a class of enzymes vital to all organisms. Their main function is to unpackage an organism's genes. They are motor proteins that move directionally along a nucleic acid phosphodiester backbone, separating two annealed nucleic acid strands (i.e., DNA, RNA, or RNA-DNA hybrid) using energy derived from ATP hydrolysis. There are many helicases resulting from the great variety of processes in which strand separation must be catalyzed. Approximately 1% of eukaryotic genes code for helicases. The human genome codes for 95 non-redundant helicases: 64 RNA helicases and 31 DNA helicases. Many cellular processes, such as DNA replication, transcription, translation, recombination, DNA repair, and ribosome biogenesis involve the separation of nucleic acid strands that necessitates the use of helicases.

Function

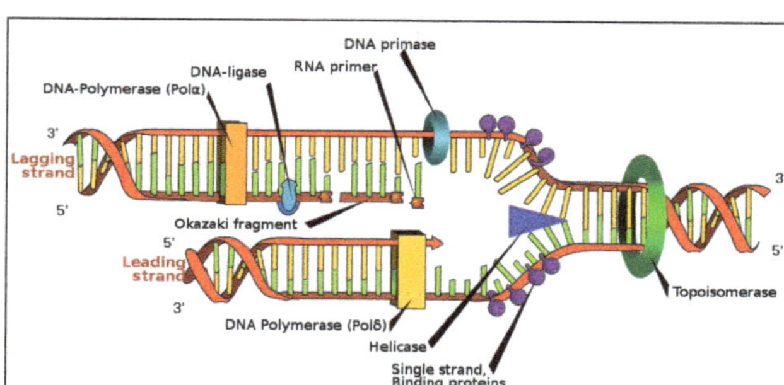

Helicase action in DNA replication.

Helicases are often used to separate strands of a DNA double helix or a self-annealed RNA molecule using the energy from ATP hydrolysis, a process characterized by the breaking of hydrogen bonds between annealed nucleotide bases. They also function to remove nucleic acid-associated proteins and catalyze homologous DNA recombination. Metabolic processes of RNA such as translation, transcription, ribosome biogenesis, RNA splicing, RNA transport, RNA editing, and RNA degradation are all facilitated by helicases. Helicases move incrementally along one nucleic acid strand of the duplex with a directionality and processivity specific to each particular enzyme.

Helicases adopt different structures and oligomerization states. Whereas DnaB-like helicases unwind DNA as ring-shaped hexamers, other enzymes have been shown to be active as monomers or dimers. Studies have shown that helicases may act passively, waiting for uncatalyzed unwinding to take place and then translocating between displaced strands, or can play an active role in catalyzing strand separation using the energy generated in ATP hydrolysis. In the latter case, the helicase acts comparably to an active motor, unwinding and translocating along its substrate as a direct result of its ATPase activity. Helicases may process much faster *in vivo* than *in vitro* due to the presence of accessory proteins that aid in the destabilization of the fork junction.

Activation Barrier in Helicase Activity

Enzymatic helicase action, such as unwinding nucleic acids is achieved through the lowering of the activation barrier (*B*) of each specific action. The activation barrier is a result of various factors, and can be defined using the following equation, where

N = number of unwound base pairs (bps),

ΔG_{bp} = free energy of base pair formation,

G_{int} = reduction of free energy due to helicase, and

G_f = reduction of free energy due to unzipping forces.

$$B = N(\Delta G_{bp} - G_{int} - G_f)$$

Factors that contribute to the height of the activation barrier include: specific nucleic acid sequence of the molecule involved, the number of base pairs involved, tension present on the replication fork, and destabilization forces.

Active and Passive Helicases

The size of the activation barrier to overcome by the helicase contributes to its classification as an active or passive helicase. In passive helicases, a significant activation barrier exists (defined as $B > k_B T$, where k_B is Boltzmann's constant and T is temperature of the system). Because of this significant activation barrier, its unwinding progression is affected largely by the sequence of nucleic acids within the molecule to unwind, and the presence of destabilization forces acting on the replication fork. Certain nucleic acid combinations will decrease unwinding rates (i.e. guanine and cytosine), while various destabilizing forces can increase the unwinding rate. In passive systems, the rate of unwinding (V_{un}) is less than the rate of translocation (V_{trans}) (translocation along the single-strand nucleic acid, ssNA). Another way to view the passive helicase is its reliance on the transient unraveling of the base pairs at the replication fork to determine its rate of unwinding.

In active helicases, $B < k_B T$, where the system lacks a significant barrier, as the helicase is able to destabilize the nucleic acids, unwinding the double-helix at a constant rate, regardless of the nucleic acid sequence. In active helicases, V_{un} is approximately equal to V_{trans}. Another way to view the active helicase is its ability to directly destabilize the replication fork to promote unwinding.

Active helicases show similar behavior when acting on both double-strand nucleic acids, dsNA, or ssNA, in regards to the rates of unwinding and rates of translocation, where in both systems V_{un} and V_{trans} are approximately equal.

These two categories of helicases may also be modelled as mechanisms. In such models the passive helicases are conceptualized as Brownian ratchets, driven by thermal fluctuations and subsequent anisotropic gradients across the DNA lattice. The active helicases, in contrast, are conceptualized as stepping motors – also known as powerstroke motors – utilizing either a conformational “inch worm” or a hand-over-hand “walking” mechanism to progress. Depending upon the organism, such helix-traversing progress can occur at rotational speeds in the range of 5,000 to 10,000 R.P.M.

Structural Features

The common function of helicases accounts for the fact that they display a certain degree of amino acid sequence homology; they all possess sequence motifs located in the interior of their primary structure, involved in ATP binding, ATP hydrolysis and translocation along the nucleic acid substrate. The variable portion of the amino acid sequence is related to the specific features of each helicase.

The presence of these helicase motifs allows putative helicase activity to be attributed to a given protein, but does not necessarily confirm it as an active helicase. Conserved motifs do, however, support an evolutionary homology among enzymes. Based on these helicase motifs, a number of helicase superfamilies have been distinguished.

Superfamilies

Helicases are classified in 6 groups (superfamilies) based on their shared sequence motifs. Helicases not forming a ring structure are in superfamilies 1 and 2, and ring-forming helicases form part of superfamilies 3 to 6. Helicases are also classified as α or β depending on if they work with single or double-strand DNA; α helicases work with single-strand DNA and β helicases work with double-strand DNA. They are also classified by translocation polarity. If translocation occurs 3′-5′ the helicase is type A; if translocation occurs 5′-3′ it is type B.

- Superfamily 1 (SF1): This superfamily can be further subdivided into SF1A and SF1B helicases. In this group helicases can have either 3′-5′ (SF1A subfamily) or 5′-3′(SF1B subfamily) translocation polarity. The most known SF1A helicases are Rep and UvrD in gram-negative bacteria and PcrA helicase from gram-positive bacteria. The most known Helicases in the SF1B group are RecD and Dda helicases.
- Superfamily 2 (SF2): This is the largest group of helicases that are involved in varied cellular processes. They are characterized by the presence of nine conserved motifs: Q, I, Ia, Ib, and II through VI. This group is mainly composed of DEAD-box RNA helicases. Some other helicases included in SF2 are the RecQ-like family and the Snf2-like enzymes. Most of the SF2 helicases are type A with a few exceptions such as the XPD family.
- Superfamily 3 (SF3): Superfamily 3 consists of helicases encoded mainly by small DNA viruses and some large nucleocytoplasmic DNA viruses. They have a 3′-5′ translocation directionality, meaning that they are all type A helicases. The most known SF3 helicase is the papilloma virus E1 helicase.
- Superfamily 4 (SF4): All SF4 family helicases have a type B polarity (5′-3′). The most studied SF4 helicase is gp4 from bacteriophage T7.
- Superfamily 5 (SF5): Rho proteins conform the SF5 group.
- Superfamily 6 (SF6): They contain the core AAA+ that is not included in the SF3 classification. Some proteins in the SF6 group are: mini chromosome maintenance MCM, RuvB, RuvA, and RuvC.

Helicase Disorders and Diseases

ATRX Helicase Mutations

The *ATRX* gene encodes the ATP-dependent helicase, ATRX (also known as XH2 and XNP) of the SNF2 subgroup family, that is thought to be responsible for functions such as chromatin remodeling, gene regulation, and DNA methylation. These functions assist in prevention of apoptosis, resulting in cortical size regulation, as well as a contribution to the survival of hippocampal and cortical structures, affecting memory and learning. This helicase is located on the X chromosome, in the pericentromeric heterochromatin and binds to heterochromatin protein 1. Studies have shown that ATRX plays a role in rDNA methylation and is essential for embryonic development. Mutations have been found throughout the *ATRX* protein, with over 90% of them being located in the zinc finger and helicase domains. Mutations of ATRX can result in X-linked-alpha-thalassaemia-mental retardation (ATR-X syndrome).

Various types of mutations found in ATRX have been found to be associated with ATR-X, including most commonly single-base missense mutations, as well as nonsense, frameshift, and deletion mutations. Characteristics of ATR-X include: microcephaly, skeletal and facial abnormalities, mental retardation, genital abnormalities, seizures, limited language use and ability, and alpha-thalassemia. The phenotype seen in ATR-X suggests that the mutation of ATRX gene causes the downregulation of gene expression, such as the alpha-globin genes. It is still unknown what causes the expression of the various characteristics of ATR-X in different patients.

XPD Helicase Point Mutations

XPD (Xeroderma pigmentosum factor D, also known as protein ERCC2) is a 5′-3′, Superfamily II, ATP-dependent helicase containing iron-sulphur cluster domains. Inherited point mutations in XPD helicase have been shown to be associated with accelerated aging disorders such as Cockayne syndrome (CS) and trichothiodystrophy (TTD). Cockayne syndrome and trichothiodystrophy are both developmental disorders involving sensitivity to UV light and premature aging, and Cockayne syndrome exhibits severe mental retardation from the time of birth. The XPD helicase mutation has also been implicated in xeroderma pigmentosa (XP), a disorder characterized by sensitivity to UV light and resulting in a several 1000-fold increase in the development of skin cancer.

XPD is an essential component of the TFIIH complex, a transcription and repair factor in the cell. As part of this complex, it facilitates nucleotide excision repair by unwinding DNA. TFIIH assists in repairing damaged DNA such as sun damage. A mutation in the XPD helicase that helps form this complex and contributes to its function causes the sensitivity to sunlight seen in all three diseases, as well as the increased risk of cancer seen in XP and premature aging seen in trichothiodystrophy and Cockayne syndrome.

XPD helicase mutations leading to trichothiodystrophy are found throughout the protein in various locations involved in protein-protein interactions. This mutation results in an unstable protein due to its inability to form stabilizing interactions with other proteins at the points of mutations. This, in turn, destabilizes the entire TFIIH complex, which leads to defects with transcription and repair mechanisms of the cell.

It has been suggested that XPD helicase mutations leading to Cockayne syndrome could be the result of mutations within XPD, causing rigidity of the protein and subsequent inability to switch from repair functions to transcription functions due to a "locking" in repair mode. This could cause the helicase to cut DNA segments meant for transcription. Although current evidence points to a defect in the XPD helicase resulting in a loss of flexibility in the protein in cases of Cockayne syndrome, it is still unclear how this protein structure leads to the symptoms described in Cockayne syndrome.

In xeroderma pigmentosa, the XPD helicase mutation exists at the site of ATP or DNA binding. This results in a structurally functional helicase able to facilitate transcription, however it inhibits its function in unwinding DNA and DNA repair. The lack of cell's ability to repair mutations, such as those caused by sun damage, is the cause of the high cancer rate in xeroderma pigmentosa patients.

RecQ Family Mutations

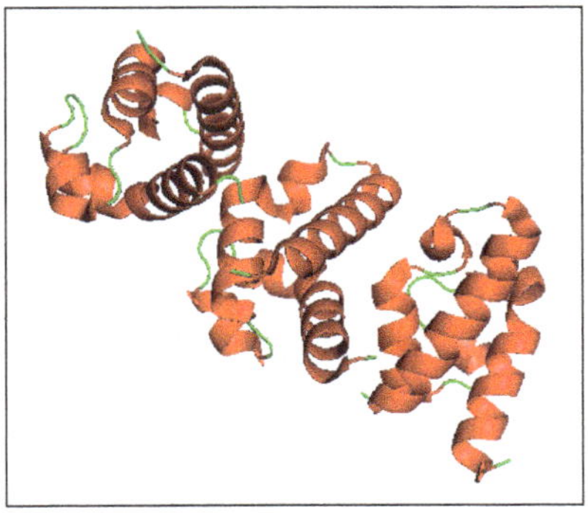

RecQ helicase

RecQ helicases (3′-5′) belong to the Superfamily II group of helicases, which help to maintain stability of the genome and suppress inappropriate recombination. Deficiencies and/or mutations in RecQ family helicases display aberrant genetic recombination and/or DNA replication, which leads to chromosomal instability and an overall decreased ability to proliferate. Mutations in RecQ family helicases BLM, RECQL4, and WRN, which play a role in regulating homologous recombination, have been shown to result in the autosomal recessive diseases Bloom syndrome (BS), Rothmund-Thomson syndrome (RTS), and Werner syndrome (WS), respectively.

Bloom syndrome is characterized by a predisposition to cancer with early onset, with a mean age-of-onset of 24 years. Cells of Bloom syndrome patients show a high frequency of reciprocal exchange between sister chromatids (SCEs) and excessive chromosomal damage. There is evidence to suggest that BLM plays a role in rescuing disrupted DNA replication at replication forks.

Werner syndrome is a disorder of premature aging, with symptoms including early onset of atherosclerosis and osteoporosis and other age related diseases, a high occurrence of sarcoma, and death often occurring from myocardial infarction or cancer in the 4th to 6th decade of life. Cells of Werner syndrome patients exhibit a reduced reproductive lifespan with chromosomal breaks and translocations, as well as large deletions of chromosomal components, causing genomic instability.

Rothmund-Thomson syndrome, also known as poikiloderma congenitale, is characterized by premature aging, skin and skeletal abnormalities, rash, poikiloderma, juvenile cataracts, and a predisposition to cancers such as osteosarcomas. Chromosomal rearrangements causing genomic instability are found in the cells of Rothmund-Thomson syndrome patients.

Meiotic Recombination

During meiosis DNA double-strand breaks and other DNA damages in a chromatid are repaired by homologous recombination using either the sister chromatid or a homologous non-sister chromatid as template. This repair can result in a crossover (CO) or, more frequently, a non-crossover (NCO) recombinant. In the yeast *Schizosaccharomyces pombe* the FANCM-family DNA helicase Fml1 directs NCO recombination formation during meiosis. The RecQ-type helicase Rqh1 also directs NCO meiotic recombination. These helicases, through their ability to unwind D-loop intermediates, promote NCO recombination by the process of synthesis-dependent strand annealing.

In the plant *Arabidopsis thaliana*, FANCM helicase promotes NCO and antagonizes the formation of CO recombinants. Another helicase, RECQ4A/B, also independently reduces COs. It was suggested that COs are restricted because of the long term costs of CO recombination, that is, the breaking up of favorable genetic combinations of alleles built up by past natural selection.

RNA Helicases

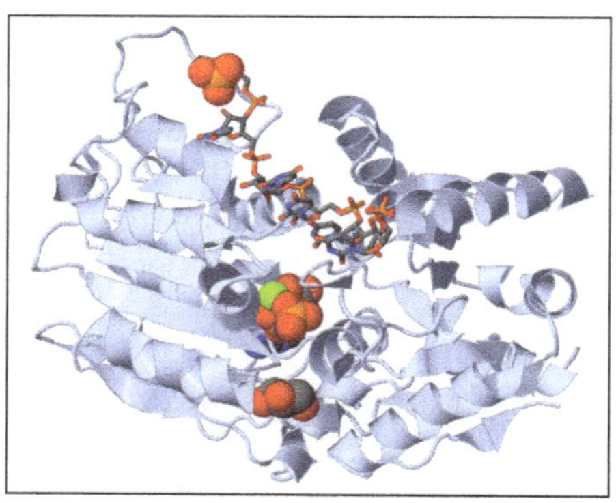

Human DEAD-box RNA helicase.

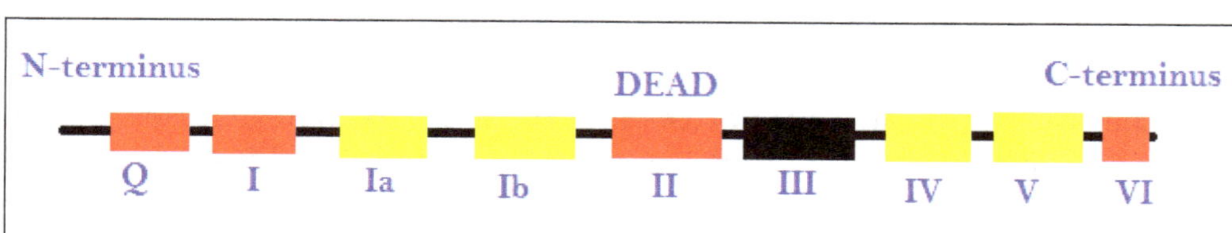

This image represents the different promoter sequences and accessory domains that aid in RNA unwinding (local strand separation). The regions in red are ATP binding domains and the regions in yellow are RNA interaction domains. Specific sequences termed DEAD box proteins are also present that help catalyze reactions in which ATP does not need to be directly hydrolyzed, as long as it binds to the domains on the strand.

RNA helicases are essential for most processes of RNA metabolism such as ribosome biogenesis, pre-mRNA splicing, and translation initiation. They also play an important role in sensing viral RNAs. RNA helicases are involved in the mediation of antiviral immune response because they can identify foreign RNAs in vertebrates. About 80% of all viruses are RNA viruses and they contain their own RNA helicases. Defective RNA helicases have been linked to cancers, infectious diseases and neuro-degenerative disorders. Some neurological disorders associated with defective RNA helicases are: amyotrophic lateral sclerosis, spinal muscular atrophy, spinocerebellar ataxia type-2, Alzheimer disease, and lethal congenital contracture syndrome.

RNA helicases and DNA helicases can be found together in all the helicase superfamilies except for SF6. All the eukaryotic RNA helicases that have been identified up to date are non-ring forming and are part of SF1 and SF2. On the other hand, ring-forming RNA helicases have been found in bacteria and viruses. However, not all RNA helicases exhibit helicase activity as defined by enzymatic function, i.e., proteins of the Swi/Snf family. Although these proteins carry the typical helicase motifs, hydrolize ATP in a nucleic acid-dependent manner, and are built around a helicase core, in general, no unwinding activity is observed.

RNA helicases that do exhibit unwinding activity have been characterized by at least two different mechanisms: canonical duplex unwinding and local strand separation. Canonical duplex unwinding is the stepwise directional separation of a duplex strand, as described above, for DNA unwinding. However, local strand separation occurs by a process wherein the helicase enzyme is loaded at any place along the duplex. This is usually aided by a single-strand region of the RNA, and the loading of the enzyme is accompanied with ATP binding. Once the helicase and ATP are bound, local strand separation occurs, which requires binding of ATP but not the actual process of ATP hydrolysis. Presented with fewer base pairs the duplex then dissociates without further assistance from the enzyme. This mode of unwinding is used by the DEAD/DEAH box helicases.

An RNA helicase database is currently available online that contains a comprehensive list of RNA helicases with information such as sequence, structure, and biochemical and cellular functions.

Diagnostic Tools for Helicase Measurement

Measuring and Monitoring Helicase Activity

Various methods are used to measure helicase activity *in vitro*. These methods range from assays that are qualitative (assays that usually entail results that do not involve values or measurements) to quantitative (assays with numerical results that can be utilized in statistical and numerical analysis). In 1982–1983, the first direct biochemical assay was developed for measuring helicase activity. This method was called a "strand displacement assay".

- Strand displacement assay involves the radiolabeling of DNA duplexes. Following helicase treatment, the single-strand DNA is visually detected as separate from the double-strand DNA by non-denaturing PAGE electrophoresis. Following detection of the single-strand DNA, the amount of radioactive tag that is on the single-strand DNA is quantified to give a numerical value for the amount of double-strand DNA unwinding.

The strand displacement assay is acceptable for qualitative analysis, its inability to display results for more than a single time point, its time consumption, and its dependence on radioactive compounds for labeling warranted the need for development of diagnostics that can monitor helicase activity in real time.

Other methods were later developed that incorporated some, if not all of the following: high-throughput mechanics, the use of non-radioactive nucleotide labeling, faster reaction time/less time consumption, real-time monitoring of helicase activity (using kinetic measurement instead of endpoint/single point analysis). These methodologies include: "a rapid quench flow method, fluorescence-based assays, filtration assays, a scintillation proximity assay, a time resolved fluorescence resonance energy transfer assay, an assay based on flashplate technology, homogenous time-resolved fluorescence quenching assays, and electrochemiluminescence-based helicase assays". With the use of specialized mathematical equations, some of these assays can be utilized to determine how many base paired nucleotides a helicase can break per hydrolysis of 1 ATP molecule.

Commercially available diagnostic kits are also available. One such kit is the "Trupoint" diagnostic assay from PerkinElmer, Inc. This assay is a time-resolved fluorescence quenching assay that utilizes the PerkinEmer "SignalClimb" technology that is based on two labels that bind in close proximity to one another but on opposite DNA strands. One label is a fluorescent lanthanide chelate, which serves as the label that is monitored through an adequate 96/384 well plate reader. The other label is an organic quencher molecule. The basis of this assay is the "quenching" or repressing of the lanthanide chelate signal by the organic quencher molecule when the two are in close proximity – as they would be when the DNA duplex is in its native state. Upon helicase activity on the duplex, the quencher and lanthanide labels get separated as the DNA is unwound. This loss in proximity negates the quenchers ability to repress the lanthanide signal, causing a detectable increase in fluorescence that is representative of the amount of unwound DNA and can be used as a quantifiable measurement of helicase activity. The execution and use of single-molecule fluorescence imaging techniques, focusing on methods that include optical trapping in conjunction with epifluorescent imaging, and also surface immobilization in conjunction with total internal reflection fluorescence visualization. Combined with microchannel flow cells and microfluidic control, allow individual fluorescently labeled protein and DNA molecules to be imaged and tracked, affording measurement of DNA unwinding and translocation at single-molecule resolution.

Determining Helicase Polarity

Helicase polarity, which is also deemed "directionality", is defined as the direction (characterized as 5′→3′ or 3′→5′) of helicase movement on the DNA/RNA single-strand along which it is moving. This determination of polarity is vital in f.ex. determining whether the tested helicase attaches to the DNA leading strand, or the DNA lagging strand. To characterize this helicase feature, a partially duplex DNA is used as the substrate that has a central single-strand DNA region with different lengths of duplex regions of DNA (one short region that runs 5′→3′ and one longer region that runs 3′→5′) on both sides of this region. Once the helicase is added to that central single-strand region, the polarity is determined by characterization on the newly formed single-strand DNA.

DNA POLYMERASE

DNA polymerase is an enzyme that synthesizes DNA molecules from deoxyribonucleotides, the building blocks of DNA. These enzymes are essential for DNA replication and usually work in pairs to create two identical DNA strands from a single original DNA molecule. During this process, DNA polymerase "reads" the existing DNA strands to create two new strands that match the existing ones.

These enzymes catalyze the following chemical reaction

$$\text{deoxynucleoside triphosphate} + \text{DNA}_n \rightleftharpoons \text{diphosphate} + \text{DNA}_{n+1}$$

DNA polymerase adds nucleotides to the three prime (3′)-end of a DNA strand, one nucleotide at a time.

Every time a cell divides, DNA polymerases are required to help duplicate the cell's DNA, so that a copy of the original DNA molecule can be passed to each daughter cell. In this way, genetic information is passed down from generation to generation.

Before replication can take place, an enzyme called helicase unwinds the DNA molecule from its tightly woven form, in the process breaking the hydrogen bonds between the nucleotide bases. This opens up or "unzips" the double-stranded DNA to give two single strands of DNA that can be used as templates for replication.

Function

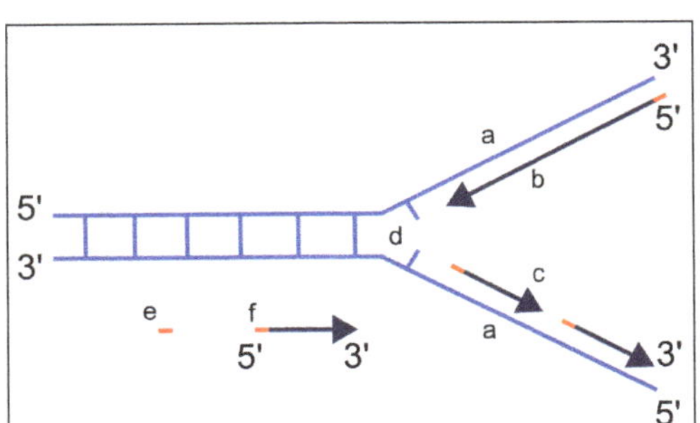

DNA polymerase moves along the old strand in the 3′–5′ direction, creating a new strand having a 5′–3′ direction.

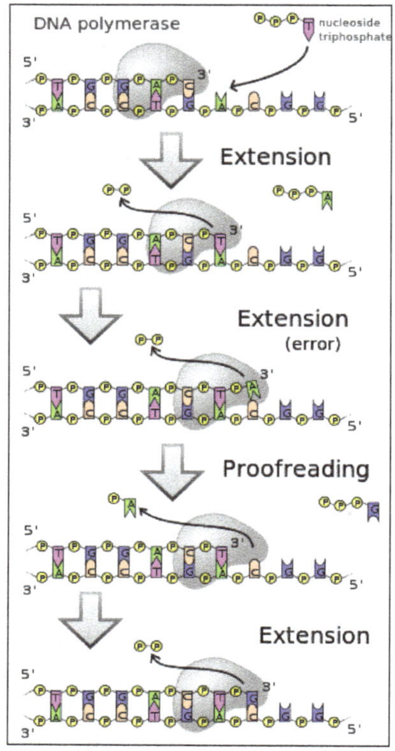

DNA polymerase with proofreading ability.

The main function of DNA polymerase is to synthesize DNA from deoxyribonucleotides, the building blocks of DNA. The DNA copies are created by the pairing of nucleotides to bases present on each strand of the original DNA molecule. This pairing always occurs in specific combinations, with cytosine along with guanine, and thymine along with adenine, forming two separate pairs, respectively. By contrast, RNA polymerases synthesize RNA from ribonucleotides from either RNA or DNA.

When synthesizing new DNA, DNA polymerase can add free nucleotides only to the 3′ end of the newly forming strand. This results in elongation of the newly forming strand in a 5′–3′ direction. No known DNA polymerase is able to begin a new chain (*de novo*); it can only add a nucleotide onto a pre-existing 3′-OH group, and therefore needs a primer at which it can add the first nucleotide. Primers consist of RNA or DNA bases (or both). In DNA replication, the first two bases are always RNA, and are synthesized by another enzyme called primase. Helicase and topoisomerase II are required to unwind DNA from a double-strand structure to a single-strand structure to facilitate replication of each strand consistent with the semiconservative model of DNA replication.

It is important to note that the directionality of the newly forming strand (the daughter strand) is opposite to the direction in which DNA polymerase moves along the template strand. Since DNA polymerase requires a free 3′ OH group for initiation of synthesis, it can synthesize in only one direction by extending the 3′ end of the preexisting nucleotide chain. Hence, DNA polymerase moves along the template strand in a 3′–5′ direction, and the daughter strand is formed in a 5′–3′ direction. This difference enables the resultant double-strand DNA formed to be composed of two DNA strands that are antiparallel to each other.

The function of DNA polymerase is not quite perfect, with the enzyme making about one mistake for every billion base pairs copied. Error correction is a property of some, but not all DNA polymerases. This process corrects mistakes in newly synthesized DNA. When an incorrect base pair is recognized, DNA polymerase moves backwards by one base pair of DNA. The 3′–5′ exonuclease activity of the enzyme allows the incorrect base pair to be excised (this activity is known as *proofreading*). Following base excision, the polymerase can re-insert the correct base and replication can continue forwards. This preserves the integrity of the original DNA strand that is passed onto the daughter cells.

Fidelity is very important in DNA replication. Mismatches in DNA base pairing can potentially result in dysfunctional proteins and could lead to cancer. Many DNA polymerases contain an exonuclease domain, which acts in detecting base pair mismatches and further performs in the removal of the incorrect nucleotide to be replaced by the correct one. The shape and the interactions accommodating the Watson and Crick base pair are what primarily contribute to the detection or error. Hydrogen bonds play a key role in base pair binding and interaction. The loss of an interaction, which occurs at a mismatch, is said to trigger a shift in the balance, for the binding of the template-primer, from the polymerase, to the exonuclease domain. In addition, an incorporation of a wrong nucleotide causes a retard in DNA polymerization. This delay gives time for the DNA to be switched from the polymerase site to the exonuclease site. Different conformational changes and loss of interaction occur at different mismatches. In a purine:pyrimidine mismatch there is a displacement of the pyrimidine towards the major groove and the purine towards the minor groove. Relative to the shape of DNA polymerase's binding pocket, steric clashes occur between the purine and residues in the minor groove, and important van der Waals and electrostatic interactions are lost by the pyrimidine. Pyrimidine:pyrimidine and purine:purine mismatches present less notable changes since the bases are displaced towards the major groove, and less steric hindrance is experienced. However, although the different mismatches result in different steric properties, DNA polymerase is still able to detect and differentiate them so uniformly and maintain fidelity in DNA replication. DNA polymerization is also critical for many mutagenesis processes and is widely employed in biotechnologies.

Structure

The known DNA polymerases have highly conserved structure, which means that their overall catalytic subunits vary very little from species to species, independent of their domain structures. Conserved structures usually indicate important, irreplaceable functions of the cell, the maintenance of which provides evolutionary advantages. The shape can be described as resembling a right hand with thumb, finger, and palm domains. The palm domain appears to function in catalyzing the transfer of phosphoryl groups in the phosphoryl transfer reaction. DNA is bound to the palm when the enzyme is active. This reaction is believed to be catalyzed by a two-metal-ion mechanism. The finger domain functions to bind the nucleoside triphosphates with the template base. The thumb domain plays a potential role in the processivity, translocation, and positioning of the DNA.

Processivity

DNA polymerase's rapid catalysis is due to its processive nature. Processivity is a characteristic of enzymes that function on polymeric substrates. In the case of DNA polymerase, the degree of processivity refers to the average number of nucleotides added each time the enzyme binds a template. The average DNA polymerase requires about one second locating and binding a primer/template junction. Once it is bound, a nonprocessive DNA polymerase adds nucleotides at a rate of one nucleotide per second. Processive DNA polymerases, however, add multiple nucleotides per second, drastically increasing the rate of DNA synthesis. The degree of processivity is directly proportional to the rate of DNA synthesis. The rate of DNA synthesis in a living cell was first determined as the rate of phage T4 DNA elongation in phage infected *E. coli*. During the period of exponential DNA increase at 37 °C, the rate was 749 nucleotides per second.

DNA polymerase's ability to slide along the DNA template allows increased processivity. There is a dramatic increase in processivity at the replication fork. This increase is facilitated by the DNA polymerase's association with proteins known as the sliding DNA clamp. The clamps are multiple protein subunits associated in the shape of a ring. Using the hydrolysis of ATP, a class of proteins known as the sliding clamp loading proteins open up the ring structure of the sliding DNA clamps allowing binding to and release from the DNA strand. Protein-protein interaction with the clamp prevents DNA polymerase from diffusing from the DNA template, thereby ensuring that the enzyme binds the same primer/template junction and continues replication. DNA polymerase changes conformation, increasing affinity to the clamp when associated with it and decreasing affinity when it completes the replication of a stretch of DNA to allow release from the clamp.

Variation Across Species

Based on sequence homology, DNA polymerases can be further subdivided into seven different families: A, B, C, D, X, Y, and RT.

Some viruses also encode special DNA polymerases, such as Hepatitis B virus DNA polymerase. These may selectively replicate viral DNA through a variety of mechanisms. Retroviruses encode an unusual DNA polymerase called reverse transcriptase, which is an RNA-dependent DNA polymerase (RdDp). It polymerizes DNA from a template of RNA.

Family	Types of DNA polymerase	Species	Examples	Feature
A	Replicative and Repair Polymerases	Eukaryotic and Prokaryotic	T7 DNA polymerase, Pol I, Pol γ, θ, and ν	Two exonuclease domains (3′-5′ and 5′-3′)
B	Replicative and Repair Polymerases	Eukaryotic and Prokaryotic	Pol II, Pol B, Pol ζ, Pol α, δ, and ε	3′-5 exonuclease (proofreading); viral ones use protein primer
C	Replicative Polymerases	Prokaryotic	Pol III	3′-5 exonuclease (proofreading)
D	Replicative Polymerases	Euryarchaeota	PolD (DP1/DP2 heterodimer)	No "hand" feature, RNA polymerase-like; 3′-5 exonuclease (proofreading)
X	Replicative and Repair Polymerases	Eukaryotic	Pol β, Pol σ, Pol λ, Pol μ, and Terminal deoxynucleotidyl transferase	template-independent; 5′ phosphatase (only Pol β)
Y	Replicative and Repair Polymerases	Eukaryotic and Prokaryotic	Pol ι, Pol κ, Pol η, Pol IV, and Pol V	
RT	Replicative and Repair Polymerases	Viruses, Retroviruses, and Eukaryotic	Telomerase, Hepatitis B virus	RNA-dependent

Prokaryotic Polymerase

Prokaryotes only have one DNA polymerase and it exists in two forms: core polymerase and holoenzyme. Core polymerase synthesizes DNA from the DNA template but it cannot initiate the synthesis alone or accurately. Holoenzyme accurately initiates synthesis.

Pol I

Prokaryotic family A polymerases include the DNA polymerase I (Pol I) enzyme, which is encoded by the *polA* gene and ubiquitous among prokaryotes. This repair polymerase is involved in excision repair with both 3′–5′ and 5′–3′ exonuclease activity and processing of Okazaki fragments generated during lagging strand synthesis. Pol I is the most abundant polymerase, accounting for >95% of polymerase activity in *E. coli*; yet cells lacking Pol I have been found suggesting Pol I activity can be replaced by the other four polymerases. Pol I adds ~15-20 nucleotides per second, thus showing poor processivity. Instead, Pol I starts adding nucleotides at the RNA primer: Template junction known as the origin of replication (ori). Approximately 400 bp downstream from the origin, the Pol III holoenzyme is assembled and takes over replication at a highly processive speed and nature.

Taq polymerase is a heat-stable enzyme of this family that lacks proofreading ability.

Pol II

DNA polymerase II is a family B polymerase encoded by the polB gene. Pol II has 3′–5′ exonuclease activity and participates in DNA repair, replication restart to bypass lesions, and its cell presence can jump from ~30-50 copies per cell to ~200–300 during SOS induction. Pol II is also thought to be a backup to Pol III as it can interact with holoenzyme proteins and assume a high level of processivity. The main role of Pol II is thought to be the ability to direct polymerase activity at the replication fork and helped stalled Pol III bypass terminal mismatches.

Pfu DNA polymerase is a heat-stable enzyme of this family found in the hyperthermophilic archaeon *Pyrococcus furiosus*. Detailed classification divides family B in archaea into B1, B2, B3, in which B2 is a group of pseudoenzymes. *Pfu* belongs to family B3. Others PolBs found in archaea are part of "Casposons", Cas1-dependent transposons. Some viruses (including Φ29 DNA polymerase) and mitochondrial plasmids carry polB as well.

Pol III

DNA polymerase III holoenzyme is the primary enzyme involved in DNA replication in *E. coli* and belongs to family C polymerases. It consists of three assemblies: the pol III core, the beta sliding clamp processivity factor, and the clamp-loading complex. The core consists of three subunits: α, the polymerase activity hub, ε, exonucleolytic proofreader, and θ, which may act as a stabilizer for ε. The beta sliding clamp processivity factor is also present in duplicate, one for each core, to create a clamp that encloses DNA allowing for high processivity. The third assembly is a seven-subunit (τ2γδδ′χψ) clamp loader complex. Recent research has classified Family C polymerases as a subcategory of Family X with no eukaryotic equivalents.

The old textbook "trombone model" depicts an elongation complex with two equivalents of the core enzyme at each replication fork (RF), one for each strand, the lagging and leading. However, recent evidence from single-molecule studies indicates an average of three stoichiometric equivalents of core enzyme at each RF for both Pol III and its counterpart in *B. subtilis,* PolC. In-cell fluorescent microscopy has revealed that leading strand synthesis may not be completely continuous, and Pol III* (i.e., the holoenzyme α, ε, τ, δ and χ subunits without the ß2 sliding clamp) has a high frequency of dissociation from active RFs. In these studies, the replication fork turnover rate was about 10s for Pol III*, 47s for the ß2 sliding clamp, and 15m for the DnaB helicase. This suggests that the DnaB helicase may remain stably associated at RFs and serve as a nucleation point for the competent holoenzyme. *In vitro* single-molecule studies have shown that Pol III* has a high rate of RF turnover when in excess, but remains stably associated with replication forks when concentration is limiting. Another single-molecule study showed that DnaB helicase activity and strand elongation can proceed with decoupled, stochastic kinetics.

Pol IV

In *E. coli*, DNA polymerase IV (Pol IV) is an error-prone DNA polymerase involved in non-targeted mutagenesis. Pol IV is a Family Y polymerase expressed by the dinB gene that is switched on via SOS induction caused by stalled polymerases at the replication fork. During SOS induction, Pol IV production is increased tenfold and one of the functions during this time is to interfere with Pol III holoenzyme processivity. This creates a checkpoint, stops replication, and allows time to repair DNA lesions via the appropriate repair pathway. Another function of Pol IV is to perform translesion synthesis at the stalled replication fork like, for example, bypassing N2-deoxyguanine adducts at a faster rate than transversing undamaged DNA. Cells lacking dinB gene have a higher rate of mutagenesis caused by DNA damaging agents.

Pol V

DNA polymerase V (Pol V) is a Y-family DNA polymerase that is involved in SOS response and translesion synthesis DNA repair mechanisms. Transcription of Pol V via the umuDC genes is highly regulated to produce only Pol V when damaged DNA is present in the cell generating an SOS response. Stalled polymerases causes RecA to bind to the ssDNA, which causes the LexA protein to autodigest. LexA then loses its ability to repress the transcription of the umuDC operon. The same RecA-ssDNA nucleoprotein posttranslationally modifies the UmuD protein into UmuD' protein. UmuD and UmuD' form a heterodimer that interacts with UmuC, which in turn activates umuC's polymerase catalytic activity on damaged DNA. In E. coli, a polymerase "tool belt" model for switching pol III with pol IV at a stalled replication fork, where both polymerases bind simultaneously to the β-clamp, has been proposed. However, the involvement of more than one TLS polymerase working in succession to bypass a lesion has not yet been shown in E. coli. Moreover, Pol IV can catalyze both insertion and extension with high efficiency, whereas pol V is considered the major SOS TLS polymerase. One example is the bypass of intra strand guanine thymine cross-link where it was shown on the basis of the difference in the mutational signatures of the two polymerases, that pol IV and pol V compete for TLS of the intra-strand crosslink.

Family D

In 1998, the family D of DNA polymerase was discovered in *Pyrococcus furiosus* and *Methanococcus jannaschii*. The PolD complex is a heterodimer of two chains, each encoded by DP1 (small proofreading) and DP2 (large catalytic). Unlike other DNA polymerases, the structure and mechanism of the catalytic core resemble that of multi-subunit RNA polymerases. The DP1-DP2 interface resembles that of Eukaryotic Class B polymerase zinc finger and its small subunit. DP1, a Mre11-like exonuclease, is likely the precursor of small subunit of Pol α and ε, providing proofreading capablities now lost in Eukaryotes. Its N-terminal HSH domain is similar to AAA proteins, especially Pol III subunit δ and RuvB, in structure. DP2 has a Class II KH domain. *Pyrococcus abyssi* polD is more heat-stable and more accurate than *Taq* polymerase, but has not yet been commercialized.

Eukaryotic DNA polymerase

Polymerases β, λ, σ and μ (Beta, Lambda, Sigma and Mu)

Family X polymerases contain the well-known eukaryotic polymerase pol β (beta), as well as other eukaryotic polymerases such as Pol σ (sigma), Pol λ (lambda), Pol μ (mu), and Terminal deoxynucleotidyl transferase (TdT). Family X polymerases are found mainly in vertebrates, and a few are found in plants and fungi. These polymerases have highly conserved regions that include two helix-hairpin-helix motifs that are imperative in the DNA-polymerase interactions. One motif is located in the 8 kDa domain that interacts with downstream DNA and one motif is located in the thumb domain that interacts with the primer strand. Pol β, encoded by POLB gene, is required for short-patch base excision repair, a DNA repair pathway that is essential for repairing alkylated or oxidized bases as well as abasic sites. Pol λ and Pol μ, encoded by the POLL and POLM genes respectively, are involved in non-homologous end-joining, a mechanism for rejoining DNA double-strand breaks due to hydrogen peroxide and ionizing radiation, respectively. TdT is expressed only in lymphoid tissue, and adds "n nucleotides" to double-strand breaks formed during V(D)J recombination to promote immunological diversity.

Polymerases α, δ and ε (Alpha, Delta and Epsilon)

Pol α (alpha), Pol δ (delta), and Pol ε (epsilon) are members of Family B Polymerases and are the main polymerases involved with nuclear DNA replication. Pol α complex (pol α-DNA primase complex) consists of four subunits: the catalytic subunit POLA1, the regulatory subunit POLA2, and the small and the large primase subunits PRIM1 and PRIM2 respectively. Once primase has created the RNA primer, Pol α starts replication elongating the primer with ~20 nucleotides. Due to its high processivity, Pol δ takes over the leading and lagging strand synthesis from Pol α. Pol δ is expressed by genes POLD1, creating the catalytic subunit, POLD2, POLD3, and POLD4 creating the other subunits that interact with Proliferating Cell Nuclear Antigen (PCNA), which is a DNA clamp that allows Pol δ to possess processivity. Pol ε is encoded by the POLE1, the catalytic subunit, POLE2, and POLE3 gene. It has been reported that the function of Pol ε is to extend the leading strand during replication, while Pol δ primarily replicates the lagging strand; however, recent evidence suggested that Pol δ might have a role in replicating the leading strand of DNA as well. Pol ε's C-terminus "polymerase relic" region, despite being unnecessary for polymerase activity, is thought to be essential to cell vitality. The C-terminus region is thought to provide a checkpoint before entering anaphase, provide stability to the holoenzyme, and add proteins to the holoenzyme necessary for initiation of replication. Pol ε has a larger "palm" domain that provides high processivity independently of PCNA.

Compared to other Family B polymerases, the DEDD exonuclease family responsible for proofreading is inactivated in Pol α. Pol ε is unique in that it has two zinc finger domains and an inactive copy of another family B polymerase in its C-terminal. The presence of this zinc finger has implications in the origins of Eukaryota, which in this case is placed into the *Asgard* group with archaeal B3 polymerase.

Polymerases η, ι and κ (Eta, Iota and Kappa)

Pol η (eta), Pol ι (iota), and Pol κ (kappa), are Family Y DNA polymerases involved in the DNA repair by translesion synthesis and encoded by genes POLH, POLI, and POLK respectively. Members of Family Y have five common motifs to aid in binding the substrate and primer terminus and they all include the typical right hand thumb, palm and finger domains with added domains like little finger (LF), polymerase-associated domain (PAD), or wrist. The active site, however, differs between family members due to the different lesions being repaired. Polymerases in Family Y are low-fidelity polymerases, but have been proven to do more good than harm as mutations that affect the polymerase can cause various diseases, such as skin cancer and Xeroderma Pigmentosum Variant (XPS). The importance of these polymerases is evidenced by the fact that gene encoding DNA polymerase η is referred as XPV, because loss of this gene results in the disease Xeroderma Pigmentosum Variant. Pol η is particularly important for allowing accurate translesion synthesis of DNA damage resulting from ultraviolet radiation. The functionality of Pol κ is not completely understood, but researchers have found two probable functions. Pol κ is thought to act as an extender or an inserter of a specific base at certain DNA lesions. All three translesion synthesis polymerases, along with Rev1, are recruited to damaged lesions via stalled replicative DNA polymerases. There are two pathways of damage repair leading researchers to conclude that the chosen pathway depends on which strand contains the damage, the leading or lagging strand.

Polymerases Rev1 and ζ (Zeta)

Pol ζ another B family polymerase, is made of two subunits Rev3, the catalytic subunit, and Rev7 (MAD2L2), which increases the catalytic function of the polymerase, and is involved in translesion synthesis. Pol ζ lacks 3′ to 5′ exonuclease activity, is unique in that it can extend primers with terminal mismatches. Rev1 has three regions of interest in the BRCT domain, ubiquitin-binding domain, and C-terminal domain and has dCMP transferase ability, which adds deoxycytidine opposite lesions that would stall replicative polymerases Pol δ and Pol ε. These stalled polymerases activate ubiquitin complexes that in turn disassociate replication polymerases and recruit Pol ζ and Rev1. Together Pol ζ and Rev1 add deoxycytidine and Pol ζ extends past the lesion. Through a yet undetermined process, Pol ζ disassociates and replication polymerases reassociate and continue replication. Pol ζ and Rev1 are not required for replication, but loss of REV3 gene in budding yeast can cause increased sensitivity to DNA-damaging agents due to collapse of replication forks where replication polymerases have stalled.

Telomerase

Telomerase is a ribonucleoprotein recruited to replicate ends of linear chromosomes because normal DNA polymerase cannot replicate the ends, or telomere. The single-strand 3′ overhang of the double-strand chromosome with the sequence 5′-TTAGGG-3′ recruits telomerase. Telomerase acts like other DNA polymerases by extending the 3′ end, but, unlike other DNA polymerases, telomerase does not require a template. The TERT subunit, an example of a reverse transcriptase, uses the RNA subunit to form the primer–template junction that allows telomerase to extend the 3′ end of chromosome ends. The gradual decrease in size of telomeres as the result of many replications over a lifetime are thought to be associated with the effects of aging.

Polymerases γ, θ and ν (Gamma, Theta and Nu)

Pol γ (gamma), Pol θ (theta), and Pol ν (nu) are Family A polymerases. Pol γ, encoded by the POLG gene, is the only mtDNA polymerase and therefore replicates, repairs, and has proofreading 3′–5′ exonuclease and 5′ dRP lyase activities. Any mutation that leads to limited or non-functioning Pol γ has a significant effect on mtDNA and is the most common cause of autosomal inherited mitochondrial disorders. Pol γ contains a C-terminus polymerase domain and an N-terminus 3′–5′ exonuclease domain that are connected via the linker region, which binds the accessory subunit. The accessory subunit binds DNA and is required for processivity of Pol γ. Point mutation A467T in the linker region is responsible for more than one-third of all Pol γ-associated mitochondrial disorders. While many homologs of Pol θ, encoded by the POLQ gene, are found in eukaryotes, its function is not clearly understood. The sequence of amino acids in the C-terminus is what classifies Pol θ as Family A polymerase, although the error rate for Pol θ is more closely related to Family Y polymerases. Pol θ extends mismatched primer termini and can bypass abasic sites by adding a nucleotide. It also has Deoxyribophosphodiesterase (dRPase) activity in the polymerase domain and can show ATPase activity in close proximity to ssDNA. Pol ν (nu) is considered to be the least effective of the polymerase enzymes. However, DNA polymerase nu plays an active role in homology repair during cellular responses to crosslinks, fulfilling its role in a complex with helicase.

Plants use two Family A polymerases to copy both the mitochrondrial and plastid genomes. They are more similar to bacterial Pol I than they are to mamallian Pol γ.

Reverse Transcriptase

Retroviruses encode an unusual DNA polymerase called reverse transcriptase, which is an RNA-dependent DNA polymerase (RdDp) that synthesizes DNA from a template of RNA. The reverse transcriptase family contain both DNA polymerase functionality and RNase H functionality, which degrades RNA base-paired to DNA. An example of a retrovirus is HIV.

TRANSFERASE

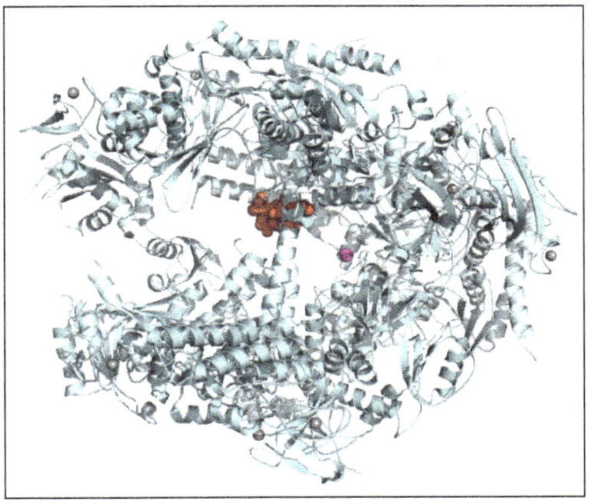

RNA polymerase from *Saccharomyces cerevisiae* complexed with α-amanitin (in red). Despite the use of the term "polymerase," RNA polymerases are classified as a form of nucleotidyl transferase.

A transferase is any one of a class of enzymes that enact the transfer of specific functional groups (e.g. a methyl or glycosyl group) from one molecule (called the donor) to another (called the acceptor). They are involved in hundreds of different biochemical pathways throughout biology, and are integral to some of life's most important processes.

Transferases are involved in myriad reactions in the cell. Three examples of these reactions are the activity of coenzyme A (CoA) transferase, which transfers thiol esters, the action of N-acetyltransferase, which is part of the pathway that metabolizes tryptophan, and the regulation of pyruvate dehydrogenase (PDH), which converts pyruvate to acetyl CoA. Transferases are also utilized during translation. In this case, an amino acid chain is the functional group transferred by a peptidyl transferase. The transfer involves the removal of the growing amino acid chain from the tRNA molecule in the A-site of the ribosome and its subsequent addition to the amino acid attached to the tRNA in the P-site.

Mechanistically, an enzyme that catalyzed the following reaction would be a transferase:

$$X\ group + Y \xrightarrow[transferase]{} X + Y\ group$$

In the above reaction, X would be the donor, and Y would be the acceptor. "Group" would be the functional group transferred as a result of transferase activity. The donor is often a coenzyme.

Nomenclature

Systematic names of transferases are constructed in the form of "donor: acceptor grouptransferase." For example, methylamine: L-glutamate N-methyltransferase would be the standard naming convention for the transferase methylamine-glutamate N-methyltransferase, where methylamine is the donor, L-glutamate is the acceptor, and methyltransferase is the EC category grouping. This same action by the transferase can be illustrated as follows:

$$\text{methylamine} + \text{L-glutamate} \rightleftharpoons NH_3 + \text{N-methyl-L-glutamate}$$

However, other accepted names are more frequently used for transferases, and are often formed as “acceptor grouptransferase” or “donor grouptransferase.” For example, a DNA methyltransferase is a transferase that catalyzes the transfer of a methyl group to a DNA acceptor. In practice, many molecules are not referred to using this terminology due to more prevalent common names. For example, RNA Polymerase is the modern common name for what was formerly known as RNA nucleotidyltransferase, a kind of nucleotidyl transferase that transfers nucleotides to the 3′ end of a growing RNA strand. In the EC system of classification, the accepted name for RNA Polymerase is DNA-directed RNA polymerase.

Classification

Described primarily based on the type of biochemical group transferred, transferases can be divided into ten categories (based on the EC Number classification). These categories comprise over 450 different unique enzymes. In the EC numbering system, transferases have been given a classification of EC2. Hydrogen is not considered a functional group when it comes to transferase targets; instead, hydrogen transfer is included under oxidoreductases, due to electron transfer considerations.

Classification of transferases into subclasses		
EC number	Examples	Group(s) transferred
EC 2.1	Methyltransferase and formyltransferase	Single-carbon groups
EC 2.2	Transketolase and transaldolase	Aldehyde or ketone groups
EC 2.3	Acyltransferase	Acyl groups or groups that become alkyl groups during transfer
EC 2.4	Glycosyltransferase, hexosyltransferase, and pentosyltransferase	Glycosyl groups, as well as hexoses and pentoses
EC 2.5	Riboflavin synthase and chlorophyll synthase	Alkyl or aryl groups, other than methyl groups
EC 2.6	Transaminase, and oximinotransferase	Nitrogenous groups
EC 2.7	Phosphotransferase, polymerase, and kinase	Phosphorus-containing groups; subclasses are based on the acceptor (e.G. Alcohol, carboxyl, etc.)
EC 2.8	Sulfurtransferase and sulfotransferase	Sulfur-containing groups
EC 2.9	Selenotransferase	Selenium-containing groups
EC 2.10	Molybdenumtransferase and tungstentransferase	Molybdenum or tungsten

Reactions

EC 2.1: Single Carbon Transferases

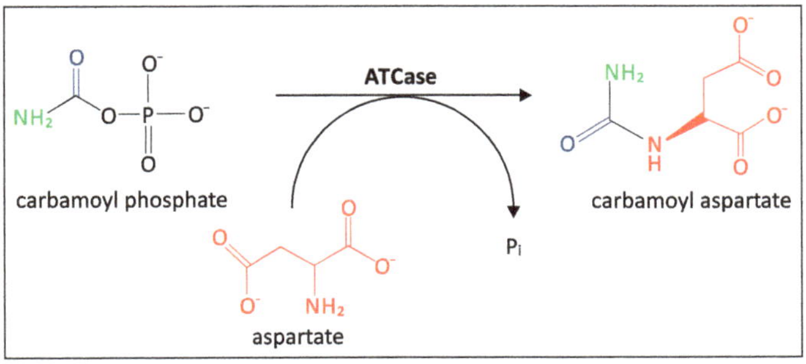

Reaction involving aspartate transcarbamylase.

EC 2.1 includes enzymes that transfer single-carbon groups. This category consists of transfers of methyl, hydroxymethyl, formyl, carboxy, carbamoyl, and amido groups. Carbamoyltransferases, as an example, transfer a carbamoyl group from one molecule to another. Carbamoyl groups follow the formula NH_2CO. In ATCase such a transfer is written as Carbamyl phosphate + L-aspertate → L-carbamyl aspartate + phosphate.

EC 2.2: Aldehyde and Ketone Transferases

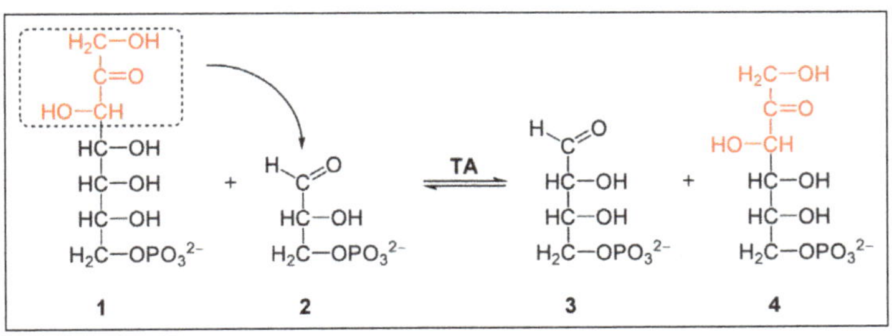

The reaction catalyzed by transaldolase.

Enzymes that transfer aldehyde or ketone groups and included in EC 2.2. This category consists of various transketolases and transaldolases. Transaldolase, the namesake of aldehyde transferases, is an important part of the pentose phosphate pathway. The reaction it catalyzes consists of a transfer of a dihydroxyacetone functional group to Glyceraldehyde 3-phosphate (also known as G3P). The reaction is as follows: sedoheptulose 7-phosphate + glyceraldehyde 3-phosphate ⇌ erythrose 4-phosphate + fructose 6-phosphate.

EC 2.3: Acyl Transferases

Transfer of acyl groups or acyl groups that become alkyl groups during the process of being transferred are key aspects of EC 2.3. Further, this category also differentiates between amino-acyl and non-amino-acyl groups. Peptidyl transferase is a ribozyme that facilitates formation of peptide bonds during translation. As an aminoacyltransferase, it catalyzes the transfer of a peptide to an aminoacyl-tRNA, following this reaction: peptidyl-$tRNA_A$ + aminoacyl-$tRNA_B$ $\rightleftharpoons$ $tRNA_A$ + peptidyl aminoacyl-$tRNA_B$.

EC 2.4: Glycosyl, Hexosyl and Pentosyl Transferases

EC 2.4 includes enzymes that transfer glycosyl groups, as well as those that transfer hexose and pentose. Glycosyltransferase is a subcategory of EC 2.4 transferases that is involved in biosynthesis of disaccharides and polysaccharides through transfer of monosaccharides to other molecules. An example of a prominent glycosyltransferase is lactose synthase which is a dimer possessing two protein subunits. Its primary action is to produce lactose from glucose and UDP-galactose. This occurs via the following pathway: UDP-β-D-galactose + D-glucose $\rightleftharpoons$ UDP + lactose.

EC 2.5: Alkyl and Aryl Transferases

EC 2.5 relates to enzymes that transfer alkyl or aryl groups, but does not include methyl groups. This is in contrast to functional groups that become alkyl groups when transferred, as those are included in EC 2.3. EC 2.5 currently only possesses one sub-class: Alkyl and aryl transferases. Cysteine synthase, for example, catalyzes the formation of acetic acids and cysteine from O_3-acetyl-L-serine and hydrogen sulfide: O_3-acetyl-L-serine + H_2S $\rightleftharpoons$ L-cysteine + acetate.

EC 2.6: Nitrogenous Transferases

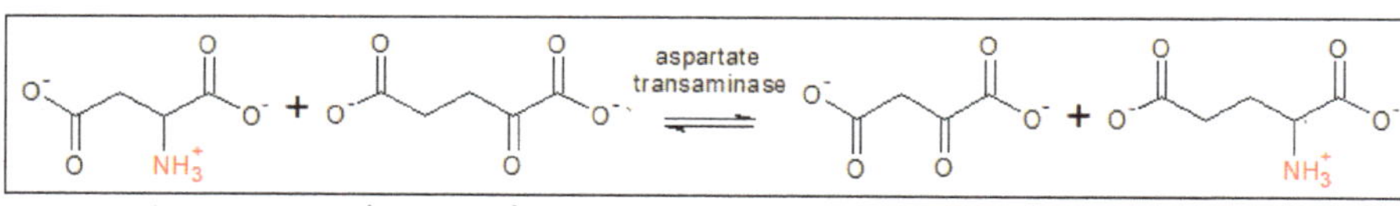

Aspartate aminotransferase can act on several different amino acids.

The grouping consistent with transfer of nitrogenous groups is EC 2.6. This includes enzymes like transaminase (also known as "aminotransferase"), and a very small number of oximinotransferases and other nitrogen group transferring enzymes. EC 2.6 previously included amidinotransferase but it has since been reclassified as a subcategory of EC 2.1 (single-carbon transferring enzymes). In the case of aspartate transaminase, which can act on tyrosine, phenylalanine, and tryptophan, it reversibly transfers an amino group from one molecule to the other.

The reaction, for example, follows the following order: L-aspartate +2-oxoglutarate oxaloacetate + L-glutamate.

EC 2.7: Phosphorus Transferases

While EC 2.7 includes enzymes that transfer phosphorus-containing groups, it also includes nuclotidyl transferases as well. Sub-category phosphotransferase is divided up in categories based on the type of group that accepts the transfer. Groups that are classified as phosphate acceptors include: alcohols, carboxy groups, nitrogenous groups, and phosphate groups. Further constituents of this subclass of transferases are various kinases. A prominent kinase is cyclin-dependent kinase (or CDK), which comprises a sub-family of protein kinases. As their name implies, CDKs are heavily dependent on specific cyclin molecules for activation. Once combined, the CDK-cyclin complex is capable of enacting its function within the cell cycle.

The reaction catalyzed by CDK is as follows: ATP + a target protein ADP + a phosphoprotein.

EC 2.8: Sulfur Transferases

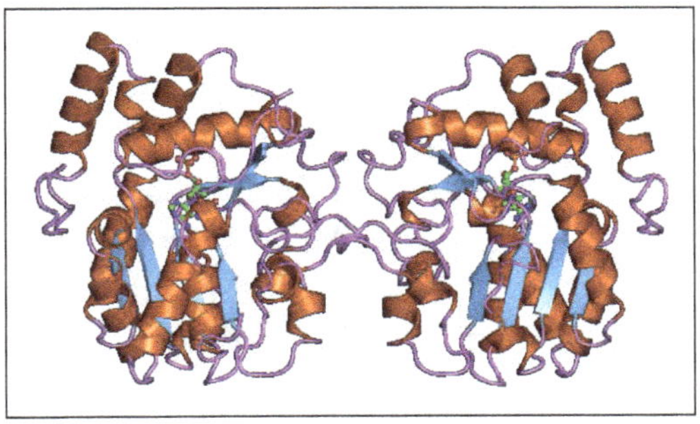

Ribbon diagram of a variant structure of estrogen sulfotransferase (PDB 1aqy EBI).

Transfer of sulfur-containing groups is covered by EC 2.8 and is subdivided into the subcategories of sulfurtransferases, sulfotransferases, and CoA-transferases, as well as enzymes that transfer alkylthio groups. A specific group of sulfotransferases are those that use PAPS as a sulfate group donor. Within this group is alcohol sulfotransferase which has a broad targeting capacity. Due to this, alcohol sulfotransferase is also known by several other names including "hydroxysteroid sulfotransferase," "steroid sulfokinase," and "estrogen sulfotransferase." Decreases in its activity has been linked to human liver disease. This transferase acts via the following reaction: 3′-phosphoadenylyl sulfate + an alcohol adenosine 3′, 5′ bisphosphate + an alkyl sulfate.

EC 2.9: Selenium Transferases

EC 2.9 includes enzymes that transfer selenium-containing groups. This category only contains two transferases, and thus is one of the smallest categories of transferase. Selenocysteine synthase, which was first added to the classification system in 1999, converts seryl-tRNA(Sec UCA) into selenocysteyl-tRNA(Sec UCA).

EC 2.10: Metal Transferases

The category of EC 2.10 includes enzymes that transfer molybdenum or tungsten-containing groups. However, as of 2011, only one enzyme has been added: molybdopterin molybdotransferase. This enzyme is a component of MoCo biosynthesis in *Escherichia coli*. The reaction it catalyzes is as follows: adenylyl-molybdopterin + molybdate molybdenum cofactor + AMP.

Role in Histo-blood Group

The A and B transferases are the foundation of the human ABO blood group system. Both A and B transferases are glycosyltransferases, meaning they transfer a sugar molecule onto an H-antigen. This allows H-antigen to synthesize the glycoprotein and glycolipid conjugates that are known as the A/B antigens. The full name of A transferase is alpha 1-3-N-acetylgalactosaminyltransferase and its function in the cell is to add N-acetylgalactosamine to H-antigen, creating A-antigen. The full name of B transferase is alpha 1-3-galactosyltransferase, and its function in the cell is to add a galactose molecule to H-antigen, creating B-antigen.

It is possible for *Homo sapiens* to have any of four different blood types: Type A (express A antigens), Type B (express B antigens), Type AB (express both A and B antigens) and Type O (express neither A nor B antigens). The gene for A and B transferases is located on chromosome 9. The gene contains seven exons and six introns and the gene itself is over 18kb long. The alleles for A and B transferases are extremely similar. The resulting enzymes only differ in 4 amino acid residues. The differing residues are located at positions 176, 235, 266, and 268 in the enzymes.

Deficiencies

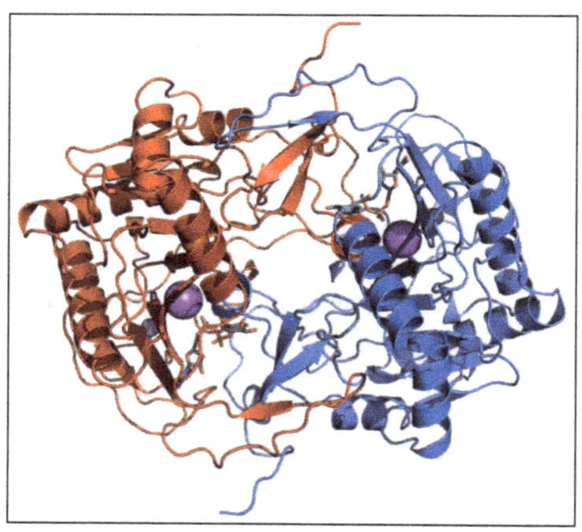

A deficiency of this transferase, E. coli galactose-1-phosphate uridyltransferase is a known cause of galactosemia.

Transferase deficiencies are at the root of many common illnesses. The most common result of a transferase deficiency is a buildup of a cellular product.

SCOT Deficiency

Succinyl-CoA:3-ketoacid CoA transferase deficiency (or SCOT deficiency) leads to a buildup of ketones. Ketones are created upon the breakdown of fats in the body and are an important energy source. Inability to utilize ketones leads to intermittent ketoacidosis, which usually first manifests during infancy. Disease sufferers experience nausea, vomiting, inability to feed, and breathing difficulties. In extreme cases, ketoacidosis can lead to coma and death. The deficiency is caused by mutation in the gene OXCT1. Treatments mostly rely on controlling the diet of the patient.

CPT-II Deficiency

Carnitine palmitoyltransferase II deficiency (also known as CPT-II deficiency) leads to an excess long chain fatty acids, as the body lacks the ability to transport fatty acids into the mitochondria to be processed as a fuel source. The disease is caused by a defect in the gene CPT2. This deficiency will present in patients in one of three ways: lethal neonatal, severe infantile hepatocardiomuscular, and myopathic form. The myopathic is the least severe form of the deficiency and can manifest at any point in the lifespan of the patient. The other two forms appear in infancy. Common symptoms of the lethal neonatal form and the severe infantile forms are liver failure, heart problems, seizures and death. The myopathic form is characterized by muscle pain and weakness following vigorous exercise. Treatment generally includes dietary modifications and carnitine supplements.

Galactosemia

Galactosemia results from an inability to process galactose, a simple sugar. This deficiency occurs when the gene for galactose-1-phosphate uridylyltransferase (GALT) has any number of mutations, leading to a deficiency in the amount of GALT produced. There are two forms of Galactosemia: classic and Duarte. Duarte galactosemia is generally less severe than classic galactosemia and is caused by a deficiency of galactokinase. Galactosemia renders infants unable to process the sugars in breast milk, which leads to vomiting and anorexia within days of birth. Most symptoms of the disease are caused by a buildup of galactose-1-phosphate in the body. Common symptoms include liver failure, sepsis, failure to grow, and mental impairment, among others. Buildup of a second toxic substance, galactitol, occurs in the lenses of the eyes, causing cataracts. Currently, the only available treatment is early diagnosis followed by adherence to a diet devoid of lactose, and prescription of antibiotics for infections that may develop.

Choline Acetyltransferase Deficiencies

Choline acetyltransferase (also known as ChAT or CAT) is an important enzyme which produces the neurotransmitter acetylcholine. Acetylcholine is involved in many neuropsychic functions such as memory, attention, sleep and arousal. The enzyme is globular in shape and consists of a single amino acid chain. ChAT functions to transfer an acetyl group from acetyl co-enzyme A to choline in the synapses of nerve cells and exists in two forms: soluble and membrane bound. The ChAT gene is located on chromosome 10.

Alzheimer's Disease

Decreased expression of ChAT is one of the hallmarks of Alzheimer's disease. Patients with Alzheimer's disease show a 30 to 90% reduction in activity in several regions of the brain, including the temporal lobe, the parietal lobe and the frontal lobe. However, ChAT deficiency is not believed to be the main cause of this disease.

Amyotrophic Lateral Sclerosis (ALS or Lou Gehrig's Disease)

Patients with ALS show a marked decrease in ChAT activity in motor neurons in the spinal cord and brain. Low levels of ChAT activity are an early indication of the disease and are detectable long before motor neurons begin to die. This can even be detected before the patient is symptomatic.

Huntington's Disease

Patients with Huntington's also show a marked decrease in ChAT production. Though the specific cause of the reduced production is not clear, it is believed that the death of medium-sized motor neurons with spiny dendrites leads to the lower levels of ChAT production.

Schizophrenia

Patients with Schizophrenia also exhibit decreased levels of ChAT, localized to the mesopontine tegment of the brain and the nucleus accumbens, which is believed to correlate with the decreased cognitive functioning experienced by these patients.

Sudden Infant Death Syndrome (SIDS)

Recent studies have shown that SIDS infants show decreased levels of ChAT in both the hypothalamus and the striatum. SIDS infants also display fewer neurons capable of producing ChAT in the vagus system. These defects in the medulla could lead to an inability to control essential autonomic functions such as the cardiovascular and respiratory systems.

Congenital Myasthenic Syndrome (CMS)

CMS is a family of diseases that are characterized by defects in neuromuscular transmission which leads to recurrent bouts of apnea (inability to breathe) that can be fatal. ChAT deficiency is implicated in myasthenia syndromes where the transition problem occurs presynaptically. These syndromes are characterized by the patients' inability to resynthesize acetylcholine.

Uses in Biotechnology

Terminal Transferases

Terminal transferases are transferases that can be used to label DNA or to produce plasmid vectors. It accomplishes both of these tasks by adding deoxynucleotides in the form of a template to the downstream end or 3′ end of an existing DNA molecule. Terminal transferase is one of the few DNA polymerases that can function without an RNA primer.

Glutathione Transferases

The family of glutathione transferases (GST) is extremely diverse, and therefore can be used for a number of biotechnological purposes. Plants use glutathione transferases as a means to segregate toxic metals from the rest of the cell. These glutathione transferases can be used to create biosensors to detect contaminants such as herbicides and insecticides. Glutathione transferases are also used in transgenic plants to increase resistance to both biotic and abiotic stress. Glutathione transferases are currently being explored as targets for anti-cancer medications due to their role in drug resistance. Further, glutathione transferase genes have been investigated due to their ability to prevent oxidative damage and have shown improved resistance in transgenic cultigens.

Rubber Transferases

Currently the only available commercial source of natural rubber is the Hevea plant (Hevea brasiliensis). Natural rubber is superior to synthetic rubber in a number of commercial uses. Efforts are being made to produce transgenic plants capable of synthesizing natural rubber, including tobacco and sunflower. These efforts are focused on sequencing the subunits of the rubber transferase enzyme complex in order to transfect these genes into other plants.

Membrane-associated Transferases

Many transferases associate with biological membranes as peripheral membrane proteins or anchored to membranes through a single transmembrane helix, for example numerous glycosyltransferases in Golgi apparatus. Some others are multi-span transmembrane proteins, for example certain Oligosaccharyltransferases or microsomal glutathione S-transferase from MAPEG family.

EXOENZYMES

An exoenzyme, or extracellular enzyme, is an enzyme that is secreted by a cell and functions outside of that cell. Exoenzymes are produced by both prokaryotic and eukaryotic cells and have been shown to be a crucial component of many biological processes. Most often these enzymes are involved in the breakdown of larger macromolecules. The breakdown of these larger macromolecules is critical for allowing their constituents to pass through the cell membrane and enter into the cell. For humans and other complex organisms, this process is best characterized by the digestive system which breaks down solid food via exoenzymes. The small molecules, generated by the exoenzyme activity, enter into cells and are utilized for various cellular functions. Bacteria and fungi also produce exoenzymes to digest nutrients in their environment, and these organisms can be used to conduct laboratory assays to identify the presence and function of such exoenzymes. Some pathogenic species also use exoenzymes as virulence factors to assist in the spread of these disease causing microorganisms. In addition to the integral roles in biological systems, different classes of microbial exoenzymes have been used by humans since pre-historic times for such diverse purposes as food production, biofuels, textile production and in the paper industry. Another important role that microbial exoenzymes serve is in the natural ecology and bioremediation of terrestrial and marine environments.

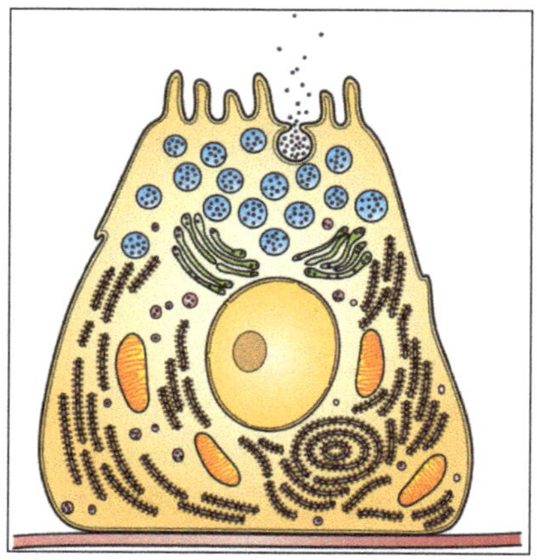

Organelles of the secretory pathway involved in the secretion of exoenzymes.

Function

In bacteria and fungi, exoenzymes play an integral role in allowing the organisms to effectively interact with their environment. Many bacteria use digestive enzymes to break down nutrients in their surroundings. Once digested, these nutrients enter the bacterium, where they are used to power cellular pathways with help from endoenzymes.

Many exoenzymes are also used as virulence factors. Pathogens, both bacterial and fungal, can use exoenzymes as a primary mechanism with which to cause disease. The metabolic activity of the exoenzymes allows the bacterium to invade host organisms by breaking down the host cells' defensive outer layers or by necrotizing body tissues of larger organisms. Many gram-negative bacteria have injectisomes, or flagella-like projections, to directly deliver the virulent exoenzyme into the host cell using a type three secretion system. With either process, pathogens can attack the host cell's structure and function, as well as its nucleic DNA.

In eukaryotic cells, exoenzymes are manufactured like any other enzyme via protein synthesis, and are transported via the secretory pathway. After moving through the rough endoplasmic reticulum, they are processed through the Golgi apparatus, where they are packaged in vesicles and released out of the cell. In humans, a majority of such exoenzymes can be found in the digestive system and are used for metabolic breakdown of macronutrients via hydrolysis. Breakdown of these nutrients allows for their incorporation into other metabolic pathways.

Examples of Exoenzymes as Virulence Factors

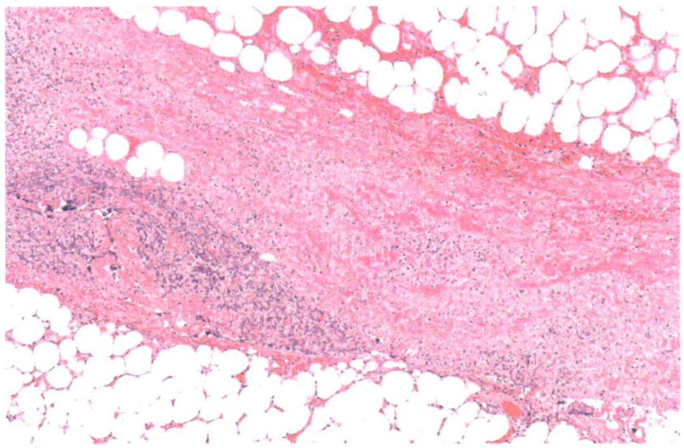

Microscopic view of necrotizing fasciitis as caused by *Streptococcus pyogenes*.

Necrotizing Enzymes

Necrotizing enzymes destroy cells and tissue. One of the best known examples is an exoenzyme produced by Streptococcus pyogenes that causes necrotizing fasciitis in humans.

Coagulase

By binding to prothrombin, coagulase facilitates clotting in a cell by ultimately converting fibrinogen to fibrin. Bacteria such as Staphylococcus aureus use the enzyme to form a layer of fibrin around their cell to protect against host defense mechanisms.

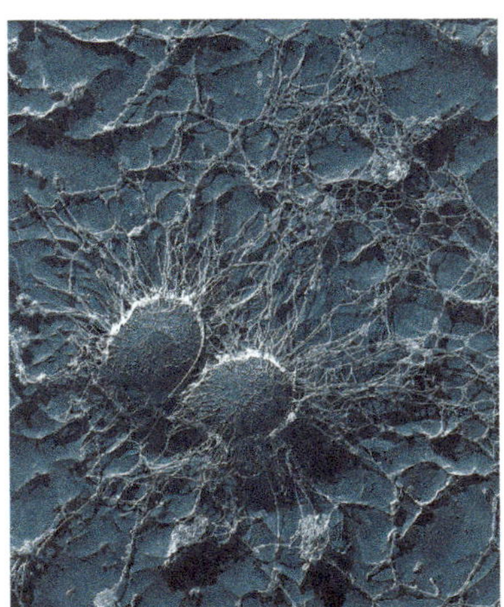

Fibrin layer formed by Staphyloccocus aureus.

Kinases

The opposite of coagulase, kinases can dissolve clots. S. aureus can also produce staphylokinase, allowing them to dissolve the clots they form, to rapidly diffuse into the host at the correct time.

Hyaluronidase

Similar to collagenase, hyaluronidase enables a pathogen to penetrate deep into tissues. Bacteria such as Clostridium do so by using the enzyme to dissolve collagen and hyaluronic acid, the protein and saccharides, respectively, that hold tissues together.

Hemolysins

Hemolysins target erythrocytes, or red blood cells. Attacking and lysing these cells allows the pathogen to harm the host organism, and also provides it with a source of iron from the lysed hemoglobin, like the fungus Candida albicans. Organisms can either by alpha-hemolytic, beta-hemolytic, or gamma-hemolytic (non-hemolytic).

Examples of Digestive Exoenzymes

Amylases

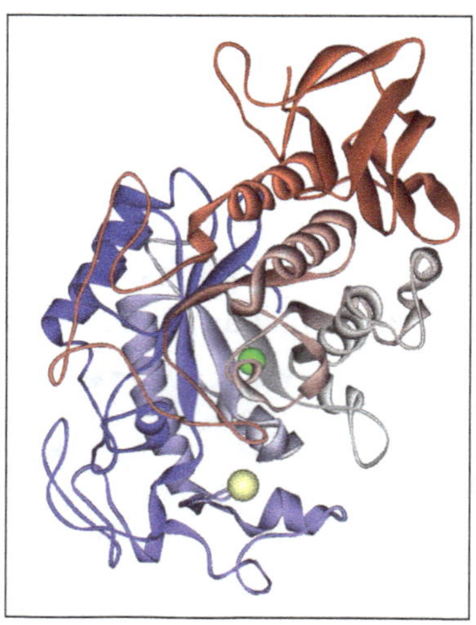

Pancreatic alpha-amylase 1HNY.

Amylases are a group of extracellular enzymes (glycoside hydrolases) that catalyze the hydrolysis of starch into maltose. These enzymes are grouped into three classes based on their amino acid sequences, mechanism of reaction, method of catalysis and their structure. The different classes of amylases are α-amylases, β-amylases, and glucoamylases. The α-amylases hydrolyze starch by randomly cleaving the 1,4-a-D-glucosidic linkages between glucose units, β-amylases cleave non-reducing chain ends of components of starch such as amylose, and glucoamylases hydrolyze glucose molecules from the ends of amylose and amylopectin. Amylases are critically important extracellular enzymes and are found in plants, animals and micro-organisms. In humans, amylases are secreted by both the pancreas and salivary glands with both sources of the enzyme required for complete starch hydrolysis.

Lipoprotein Lipase

Lipoprotein lipase (LPL) is a type of digestive enzyme that helps regulate the uptake of triacylglycerols from chylomicrons and other low-density lipoproteins from fatty tissues in the body. The exoenzymatic function allows it to break down the triacylglycerol into two free fatty acids and one molecule of monoacylglycerol. LPL can be found in endothelial cells in fatty tissues, such as adipose, cardiac, and muscle. Lipoprotein lipase is downregulated by high levels of insulin, and upregulated by high levels of glucagon and adrenaline.

Pectinase

Pectinases, also called pectolytic enzymes, are a class of exoenzymes that are involved in the breakdown of pectic substances, most notably pectin. Pectinases can be classified into two different groups based on their action against the galacturonan backbone of pectin: de-esterifying and depolymerizing. These exoenzymes can be found in both plants and microbial organisms including fungi and bacteria. Pectinases are most often used to break down the pectic elements found in plants and plant-derived products.

Pepsin

Discovered in 1836, pepsin was one of the first enzymes to be classified as an exoenzyme. The enzyme is first made in the inactive form, pepsinogen by chief cells in the lining of the stomach. With an impulse from the vagus nerve, pepsinogen is secreted into the stomach, where it mixes with hydrochloric acid to form pepsin. Once active, pepsin works to break down proteins in foods such as dairy, meat, and eggs. Pepsin works best at the pH of gastric acid, 1.5 to 2.5, and is deactivated when the acid is neutralized to a pH of 7.

Trypsin

Also one of the first exoenzymes to be discovered, trypsin was named in 1876, forty years after pepsin. This enzyme is responsible for the breakdown of large globular proteins and its activity is specific to cleaving the C-terminal sides of arginine and lysine amino acid residues. It is the derivative of trypsinogen, an inactive precursor that is produced in the pancreas. When secreted into the small intestine, it mixes with enterokinase to form active trypsin. Due to its role in the small intestine, trypsin works at an optimal pH of 8.0.

Bacterial Assays

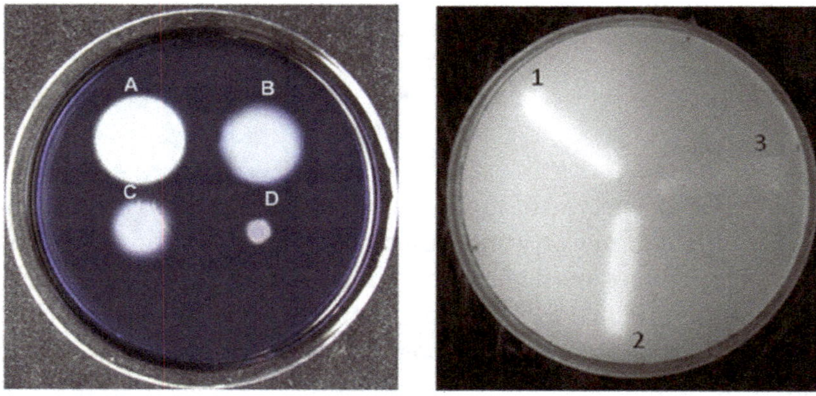

Results of bacterial assays. Left:amylase bacterial assay on a starch medium. A indicates a positive result, D indicates a negative result. Right: lipase bacterial assay on an olive oil medium. 1 shows a positive result, 3 shows a negative result.

The production of a particular digestive exoenzyme by a bacterial cell can be assessed using plate assays. Bacteria are streaked across the agar, and are left to incubate. The release of the enzyme into the surroundings of the cell cause the breakdown of the macromolecule on the plate. If a reaction does not occur, this means that the bacteria does not create an exoenzyme capable of interacting with the surroundings. If a reaction does occur, it becomes clear that the bacteria does possess an exoenzyme, and which macromolecule is hydrolyzed determines its identity.

Amylase

Amylase breaks down carbohydrates into mono- and disaccharides, so a starch agar must be used for this assay. Once the bacteria is streaked on the agar, the plate is flooded with iodine. Since iodine binds to starch but not its digested byproducts, a clear area will appear where the amylase reaction has occurred. Bacillus subtilis is a bacterium that results in a positive assay as shown in the picture.

Lipase

Lipase assays are done using a lipid agar with a spirit blue dye. If the bacteria has lipase, a clear streak will form in the agar, and the dye will fill the gap, creating a dark blue halo around the cleared area. Staphylococcus epidermis results in a positive lipase assay.

Biotechnological and Industrial Applications

Microbiological sources of exoenzymes including amylases, proteases, pectinases, lipases, xylanases, cellulases among others are used for a wide range of biotechnological and industrial uses including biofuel generation, food production, paper manufacturing, detergents and textile production. Optimizing the production of biofuels has been a focus of researchers in recent years and is centered around the use of microorganisms to convert biomass into ethanol. The enzymes that are of particular interest in ethanol production are cellobiohydrolase which solubilizes crystalline cellulose and xylanase that hydrolyzes xylan into xylose. One model of biofuel production is the use of a mixed population of bacterial strains or a consortium that work to facilitate the breakdown of cellulose materials into ethanol by secreting exoenzymes such as cellulases and laccases. In addition to the important role it plays in biofuel production, xylanase is utilized in a number of other industrial and biotechnology applications due to its ability to hydrolyze cellulose and hemicellulose. These applications include the breakdown of agricultural and forestry wastes, working as a feed additive to facilitate greater nutrient uptake by livestock, and as an ingredient in bread making to improve the rise and texture of the bread.

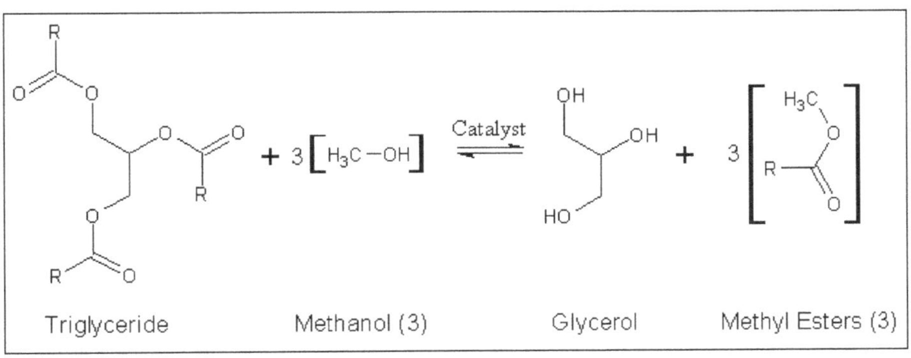

Generic Biodiesel Reaction: Lipases can serve as a biocatalyst in this reaction.

Lipases are one of the most used exoenzymes in biotechnology and industrial applications. Lipases make ideal enzymes for these applications because they are highly selective in their activity, they are readily produced and secreted by bacteria and fungi, their crystal structure is well characterized, they do not require cofactors for their enzymatic activity, and they do not catalyze side reactions. The range of uses of lipases encompasses production of biopolymers, generation of cosmetics, use as a herbicide, and as an effective solvent. However, perhaps the most well known use of lipases in this field is its use in the production of biodiesel fuel. In this role, lipases are used to convert vegetable oil to methyl- and other short-chain alcohol esters by a single transesterification reaction.

Cellulases, hemicellulases and pectinases are different exoenzymes that are involved in a wide variety of biotechnological and industrial applications. In the food industry these exoenzymes are used in the production of fruit juices, fruit nectars, fruit purees and in the extraction of olive oil among many others. The role these enzymes play in these food applications is to partially breakdown the plant cell walls and pectin. In addition to the role they play in food production, cellulases are used in the textile industry to remove excess dye from denim, soften cotton fabrics and restore the color brightness of cotton fabrics. Cellulases and hemicellulases (including xylanases) are also used in the paper and pulp industry to de-ink recycled fibers, modify coarse mechanical pulp and for the partial or complete hydrolysis of pulp fibers. Cellulases and hemicellulases are used in these industrial applications due to their ability to hydrolyze the cellulose and hemicellulose components found in these materials.

Bioremediation Applications

Water pollution from runoff of soil and fertilizer.

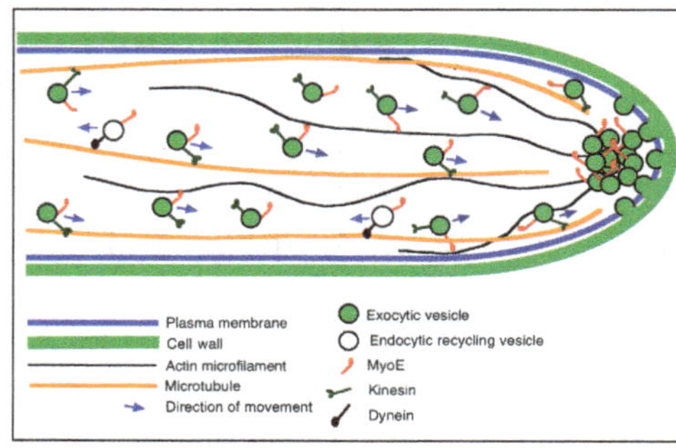

Exocytic vesicles move along actin microfilaments towards the fungal hyphal tip where they release their contents including exoenzymes.

Bioremediation is a process in which pollutants or contaminants in the environment are removed through the use of biological organisms or their products. The removal of these often hazardous pollutants is mostly carried out by naturally occurring or purposely introduced microorganisms that are capable of breaking down or absorbing the desired pollutant. The types of pollutants that are often the targets of bioremediation strategies are petroleum products (including oil and solvents) and pesticides. In addition to the microorganisms ability to digest and absorb the pollutants, their secreted exoenzymes play an important role in many bioremediation strategies.

Fungi have been shown to be viable organisms to conduct bioremediation and have been used to aid in the decontamination of a number of pollutants including polycyclic aromatic hydrocarbons (PAHs), pesticides, synthetic dyes, chlorophenols, explosives, crude oil and many others. While fungi can breakdown many of these contaminants intracellularly, they also secrete numerous oxidative exoenzymes that work extracellularly. One critical aspect of fungi in regards to bioremediation is that they secrete these oxidative exoenzymes from their ever elongating hyphal tips. Laccases are an important oxidative enzyme that fungi secrete and use oxygen to oxidize many pollutants. Some of the pollutants that laccases have been used to treat include dye-containing effluents from the textile industry, wastewater pollutants (chlorophenols, PAHs, etc.), and sulphur-containing compounds from coal processing.

Bacteria are also a viable source of exoenzymes capable to facilitating the bioremediation of the environment. There are many examples of the use of bacteria for this purpose and their exoenzymes encompass many different classes of bacterial enzymes. Of particular interest in this field are bacterial hydrolases as they have an intrinsic low substrate specificity and can be used for numerous pollutants including solid wastes. Plastic wastes including polyurethanes are particularly hard to degrade, but an exoenzyme has been identified in a Gram-negative bacterium, *Comamonas acidovorans*, that was capable of degrading polyurethane waste in the environment. Cell-free use of microbial exoenzymes as agents of bioremediation is also possible although their activity is often not as robust and introducing the enzymes into certain environments such as soil has been challenging. In addition to terrestrial based microorganisms, marine based bacteria and their exoenzymes show potential as candidates in the field of bioremediation. Marine based bacteria have been utilized in the removal of heavy metals, petroleum/diesel degradation and in the removal of polyaromatic hydrocarbons among others.

ENDOENZYME

An endoenzyme, or intracellular enzyme, is an enzyme that functions within the cell in which it was produced. Because the majority of enzymes fall within this category, the term is used primarily to differentiate a specific enzyme from an exoenzyme. It is possible for a single enzyme to have both endoenzymatic and exoenzymatic functions; for example, glycolytic enzymes of Kreb's Cycle. In most cases the term endoenzyme refers to an enzyme that binds to a bond 'within the body' of a large molecule - usually a polymer. For example an endoamylase would break down large amylose molecules into shorter dextrin chains. On the other hand, an exoenzyme removes subunits from the polymer one at a time from one end; in effect it can only act at the end ponts of a polymer. An exoamylase would therefore remove one glucose molecule at a time from the end of an amylose molecule.

ZYMOGENS

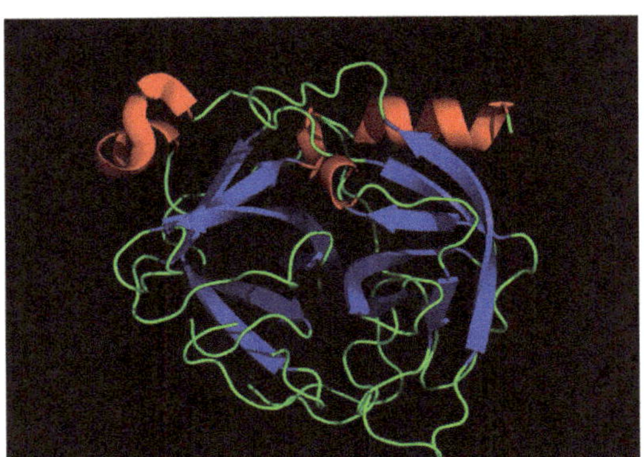

A graphical representation of zymogen.

A zymogen(also denoted as a proenzyme) is a group of proteins that can also be described as an inactive enzyme. Since it is an inactive precursor, it does not hold any catalytic activity. These zymogens can be activated by chemical processes such as cleaving, hydrolysis, along with other biochemical changes that cleave the inactive enzyme to make it active. These biochemical processes often occurs in a lysosome, where cleavage reveals the active site. Exposure of active site allow the enzyme to become active and function to catalyze reactions. The reason for cells to secrete inactive enzymes is to prevent unwanted destruction of cellular proteins. It is only when the conditions are right that zymogens become activated into enzymes. There are specialized zymogenic cells that work only to synthesize and store zymogens in inactive form, ready to send to the parts of the body at need.

Types of Zymogens

Pepsinogen

Pepsinogen, inactive precursor form of pepsin, is secreted by Chief cells in the stomach. Pepsinogen is activated by Hydrochloric acid (secretion from Parietal cells) because Hydrochloric acid provides the necessary acidic environment for which pepsin works best. Once pepsinogen becomes pepsin, it is responsible for the breakdown of food. The difference between pepsinogen and pepsin is that its primary structure has an additional 44 amino acids. The activation of pepsinogen to pepsin makes pepsin available to catalyze pepsinogen further to cleave it into more pepsin.

Trypsinogen

Trypsinogen is the inactive form of trypsin. It is secreted by the pancreas and found in pancreatic juice. Once activated in the duodenum, trypsin cleaves peptide chains at the carboxyl side of the amino acids lysine and arginine unless they are preceding a proline. The importance of trypsin is its ability to cleave other zymogens such as chymotripsinogen and procarboxypeptidase.

Chymotrypsinogen

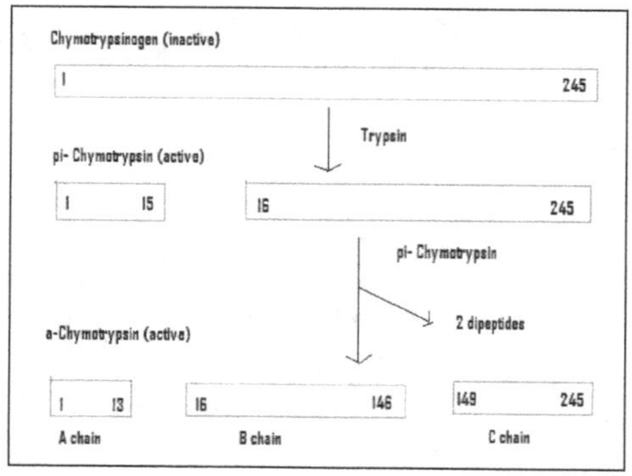

Chymatrypsinogen is a zymogen that is yet to be activated into an active enzyme. After the cleavage of peptide bonds, the inactive form is not an fully active form with 3 different chains.

Once chymotrypsinogen is cleaved by trypsin, and reacted with Chymotrypsin produces a fully active enzyme, Chymotrypsin. Chymotrypsin is found in the digestive system of mammals, as well as other organisms. It works by cleaving peptides at the carboxyl end of aromatic amino acids (Tryptophan, Tyrosin, and Phenylalanine).

Procarboxypeptidase

Procarboxypeptidase, which is the inactive form of carboxypeptidase, is converted to the active form by trypsin and enteropeptidase. It is secreted by the pancreas. Only some forms of carboxypeptidase are initially produced in the inactive form. However, the advantage of this mechanism is to ensure that the enzymes are not immediately exhausted before digestion.

Nuclease

Nuclease is an enzyme which can break phosphodiester bonds between nucleotides in a DNA sequence. Nucleases contain a general range of enzymes such as endonucleases. Nucleases vary in the DNA sequences they cut as certain phosphodiester bonds are cleaved in such a way that may not always completely symmetrical.

Pancreatic Amylase

Pancreatic Amylase is an enzyme that converts complex sugars such as starches and polysaccharides of carbohydrates into simpler sugars during digestion. Amylase hydrolyzes starch, glycogen, and dextrin to form in all three instances glucose, maltose, and the limit-dextrins. Amylase is mainly secreted by the salivary glands, but some may also be found in the pancreas that also help aid in digestion.

Lipase

Lipase is the active form of prolipase. Once activated, the water soluble enzyme is the catalyst for the hydrolysis ester bonds in water-insoluble, lipid substrates. This action is what classifies lipases as a subclass of the esterases. It cleaves fats into monoglycerides, fatty acids, and glycerols. The lipase also is essential in the process of digestion, as well as the transport and processing of dietary lipids in almost all organisms.

Proelastase

Proelastease is the inactive form of elastase. The activation of Proelastease is initially done through the simple cleavage of multiple sub-unit residues that bind to the central structure of the protein structure. The cleavage disrupts the hydrophobic interactions of the tertiary structure, allowing the polar regions of the enzyme to respond to digestion. Proleastease is distinct from Chymotrypsin through isolation by chromatography reveals that this structure is highly resistant and further stabilized by internal hydrogen bonding. Activation of proelastase and propeptidase with trypsin changed their electrophoretic mobilities. The activated proelastase migrated at the same rate as authentic, pure elastase.The proenzymes that were originally highly insoluble could be solubilized by treatment with alumina Cγ gel.

Enteropeptidase

Enteropeptidase is produced within the walls of the small intestines secreted by the duodenum glands. This enzyme proteolytically activates trypsinogen to trypsin which simultaneously activates other digestive enzymes as well. Enteropeptidase cleaves at the C-terminal end of trypsinogen which activates the enzyme, turning trypsinogen into trypsin. Most of Enteropeptidase is consisted of disulfide bonds within the larger chain forming the catalytic subunits.

Caspase

Caspases are a family of cysteine proteases that play essential roles in apoptosis, necrosis, and inflammation. Caspases plays an important role in cells for apoptosis (programmed death) during the development and other stages of adult life. Caspases are also known as "executioner" proteins because of this role that they play in the cell.

Prothrombin

Prothrombin(Zymogen) is the precursor to the enzyme Thrombin which in turn converts fibrinogen in to fibrin. Fibrin is the protein responsible for blood clotting and tissue restoration during a tissue rupture. Fibrinogen is the substrate to thrombin and when activated by thrombin it becomes Fibrin a non-soluble glycol protein. Fibrimogen has a linearly symmetrical structure, containing a central cleavage site and a at both ends it has what is known as a globular unit which comes after a designated region alpha. The globular region has 2 "connection sites" Beta and Gamma. These 2 sites have the ability to connect to 2 other fibrin proteins, this allows the proteins to connect and make a lattice like structure making it a strong structure known as a cross linked fibrin clot.

Angiotensin

Angiotensin is an oligopeptide that causes blood vessels to constrict and increased blood pressure. It also stimulates the release of aldosterone from the adrenal cortex. It is a hormone and a powerful dipsogen.

ISOENZYMES

Isozymes (also known as isoenzymes or more generally as multiple forms of enzymes) are enzymes that differ in amino acid sequence but catalyze the same chemical reaction.

These enzymes usually display different kinetic parameters (e.g. different K_M values), or different regulatory properties. The existence of isozymes permits the fine-tuning of metabolism to meet the particular needs of a given tissue or developmental stage. In biochemistry, isozymes (or isoenzymes) are isoforms (closely related variants) of enzymes. In many cases, they are coded for by homologous genes that have diverged over time. Although, strictly speaking, allozymes represent enzymes from different alleles of the same gene, and isozymes represent enzymes from different genes that process or catalyse the same reaction, the two words are usually used interchangeably.

Isozymes were first described by R. L. Hunter and Clement Markert who defined them as different variants of the same enzyme having identical functions and present in the same individual. This definition encompasses (1) enzyme variants that are the product of different genes and thus represent different loci (described as isozymes) and (2) enzymes that are the product of different alleles of the same gene (described as allozymes).

Isozymes are usually the result of gene duplication, but can also arise from polyploidisation or nucleic acid hybridization. Over evolutionary time, if the function of the new variant remains *identical* to the original, then it is likely that one or the other will be lost as mutations accumulate, resulting in a pseudogene. However, if the mutations do not immediately prevent the enzyme from functioning, but instead modify either its function, or its pattern of expression, then the two variants may both be favoured by natural selection and become specialised to different functions. For example, they may be expressed at different stages of development or in different tissues.

Allozymes may result from point mutations or from insertion-deletion (indel) events that affect the coding sequence of the gene. As with any other new mutations, there are three things that may happen to a new allozyme:

- It is most likely that the new allele will be non-functional—in which case it will probably result in low fitness and be removed from the population by natural selection.
- Alternatively, if the amino acid residue that is changed is in a relatively unimportant part of the enzyme (e.g., a long way from the active site), then the mutation may be selectively neutral and subject to genetic drift.
- In rare cases, the mutation may result in an enzyme that is more efficient, or one that can catalyse a slightly different chemical reaction, in which case the mutation may cause an increase in fitness, and be favoured by natural selection.

An example of an isozyme is glucokinase, a variant of hexokinase which is not inhibited by glucose 6-phosphate. Its different regulatory features and lower affinity for glucose (compared to other hexokinases), allow it to serve different functions in cells of specific organs, such as control of insulin release by the beta cells of the pancreas, or initiation of glycogen synthesis by liver cells. Both these processes must only occur when glucose is abundant.

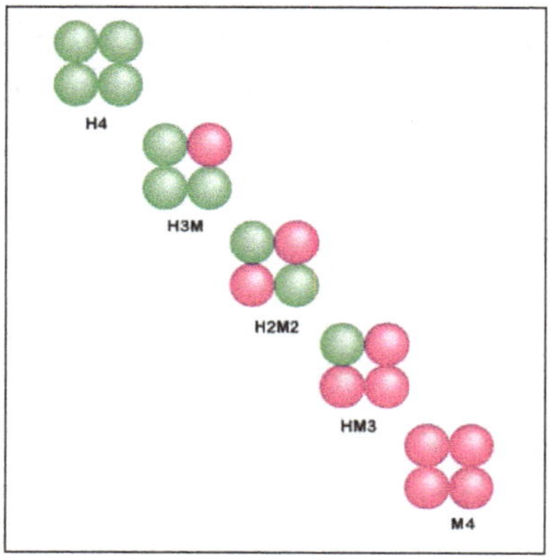

The 5 isozymes of LDH.

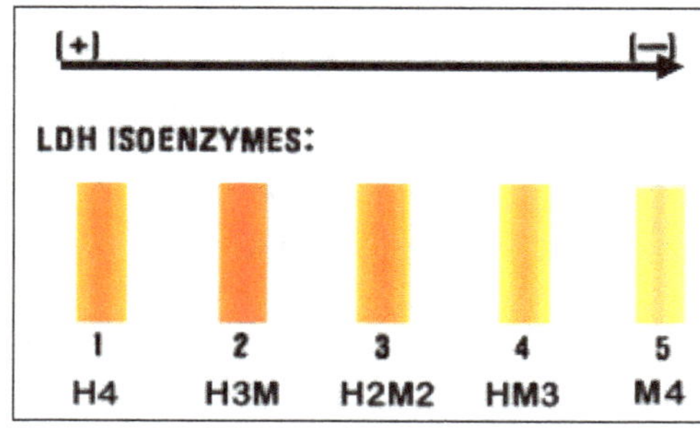

Distinction between five isozymes using electrophoresis.

1.) The enzyme lactate dehydrogenase is a tetramer made of two different sub-units, the H-form and the M-form. These combine in different combinations depending on the tissue:

Type	Composition	Location	Electrophoretic Mobility	Whether destroyed by Heat (at 60° C)	Percentage of normal serum in humans
LDH_1	HHHH	Heart and Erythrocyte	Fastest	No	25%
LDH_2	HHHM	Heart and Erythrocyte	Faster	No	35%
LDH_3	HHMM	Brain and Kidney	Fast	Partially	27%
LDH_4	HMMM	Skeletal Muscle and Liver	Slow	Yes	8%
LDH_5	MMMM	Skeletal Muscle and Liver	Slowest	Yes	5%

2.) Isoenzymes of creatine phosphokinase: Creatine kinase (CK) or creatine phosphokinase (CPK) catalyses the interconversion of phospho creatine to creatine.

CPK exists in 3 isoenzymes. Each isoenzymes is a dimer of 2 subunits M (muscle), B (brain) or both

Isoenzyme	Subunit	Tissue of Origin
CPK_1	BB	Brain
CPK_2	MB	Heart
CPK_3	MM	Skeletal muscle

3.) Isoenzymes of alkaline phosphatase: Six isoenzymes have been identified. The enzyme is a monomer, the isoenzymes are due to the differences in the carbohydrate content (sialic acid residues). The most important ALP isoenzymes are α_1- ALP, α_2- heat labile ALP, α_2- heat stable ALP, pre-β ALP and γ- ALP.Increase in α_2- heat labile ALP suggests hepatitis whereas pre-β ALP indicates bone diseases.

Distinguishing Isozymes

Isozymes (and allozymes) are variants of the same enzyme. Unless they are identical in their biochemical properties, for example their substrates and enzyme kinetics, they may be distinguished by a biochemical assay. However, such differences are usually subtle, particularly between *allozymes* which are often neutral variants. This subtlety is to be expected, because two enzymes that differ significantly in their function are unlikely to have been identified as *isozymes*.

While isozymes may be almost identical in function, they may differ in other ways. In particular, amino acid substitutions that change the electric charge of the enzyme are simple to identify by gel electrophoresis, and this forms the basis for the use of isozymes as molecular markers. To identify isozymes, a crude protein extract is made by grinding animal or plant tissue with an extraction buffer, and the components of extract are separated according to their charge by gel electrophoresis. Historically, this has usually been done using gels made from potato starch, but acrylamide gels provide better resolution.

All the proteins from the tissue are present in the gel, so that individual enzymes must be identified using an assay that links their function to a staining reaction. For example, detection can be based on the localised precipitation of soluble indicator dyes such as tetrazolium salts which become insoluble when they are reduced by cofactors such as NAD or NADP, which generated in zones of enzyme activity. This assay method requires that the enzymes are still functional after separation (native gel electrophoresis), and provides the greatest challenge to using isozymes as a laboratory technique.

21-HYDROXYLASE

Steroid 21-hydroxylase, also called steroid 21-monooxygenase, 21α-hydroxylase, P45021A2, and, less commonly 21β-hydroxylase, is a cytochrome P450 enzyme that is involved with the biosynthesis of the steroid hormones aldosterone and cortisol. These syntheses take place in the adrenal cortex. Specifically, 21-hydroxylase converts progesterone and 17α-hydroxyprogesterone into 11-deoxycorticosterone and 11-deoxycortisol, respectively, by hydroxylating at the C21 position. The products of the conversions then continue through their appropriate pathways towards creation of aldosterone and cortisol. Like other cytochrome P450 enzymes, 21-hydroxylase participates in the cytochrome P450 catalytic cycle, and engages in one-electron transfer with NADPH-P450 reductase. Its structure includes an essential iron heme group centered within the protein, also common to all P450 enzymes. Variations of the 21-hydroxylase enzyme can be found in all vertebrates. However, understanding of human 21-hydroxylase structure and function is of particular clinical value, as a failure of the enzyme to act appropriately results in congenital adrenal hyperplasia. The x-ray crystal structure for human 21-hydroxylase, with bound progesterone, was realized and published in 2015, providing opportunity for further study. The enzyme is notable for its substrate specificity and relatively high catalytic efficiency.

Structure

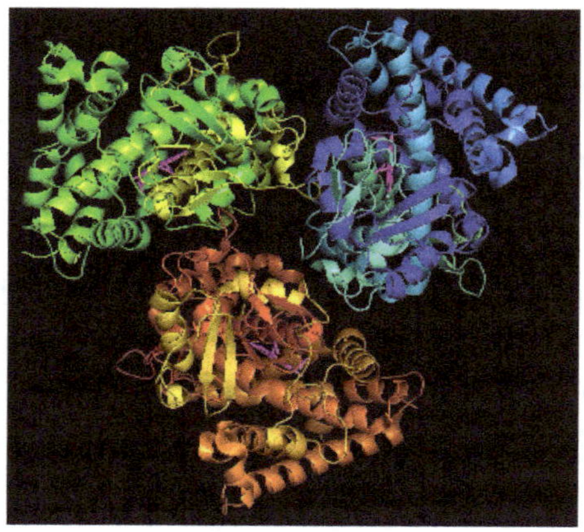

Full structure of Human 21-Hydroxylase, showing three identical subunits, each with a centralized heme group (magenta).

21-hydroxylase is a complex of three independent and identical enzyme subunits. Each subunit in the human enzyme consists of 13 α-helices and 9 ß-strands, formed into a triangular prism-like tertiary structure. The iron(III) heme group that defines the active site resides in the center of each subunit. The human enzyme binds one substrate at a time. In contrast, the well-characterized bovine enzyme can bind two substrates. The human and bovine enzyme share 80% amino acid sequence identity, but are structurally different, particularly in loop regions, and also evident in secondary structure elements.

Function

This gene encodes a member of the cytochrome P450 superfamily of enzymes. The cytochrome P450 proteins are monooxygenases which catalyze many reactions involved in drug metabolism and synthesis of cholesterol, steroids and other lipids. This protein localizes to the endoplasmic reticulum and hydroxylates steroids at the 21 position. The 21-hydroxylase enzyme is one of three microsomal steroidogenic P450 enzymes, the others being 17-hydroxylase and aromatase. 21-hydroxylase is an essential enzyme in the biosynthetic pathways that produce cortisol and aldosterone.

Reaction

21-Hydroxylase catalyzes the addition of hydroxyl (-OH) to the C21 position of two steroids: progesterone and 17α-hydroxyprogesterone.

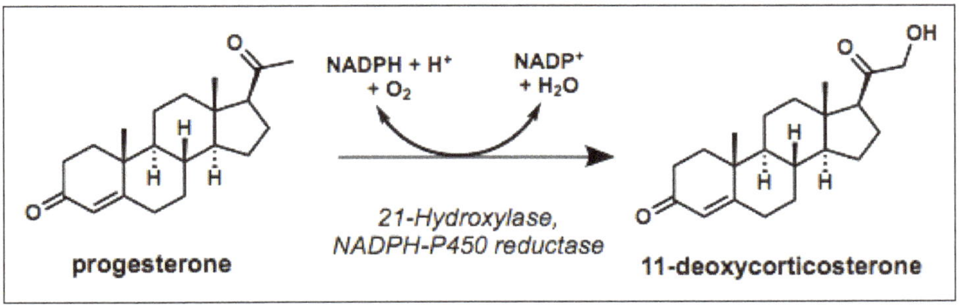

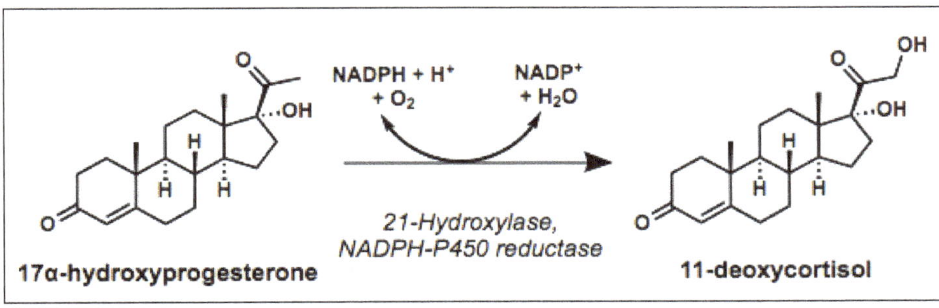

Mechanism

21-Hydroxylase is a cytochrome P450 enzyme and follows the P450 catalytic cycle.

Kinetics

21-Hydroxylase is highly specific for hydroxylation of progesterone and 17-hydroxyprogesterone. No studies have reported sufficient binding of alternate substrates. In this way, it differs from the evolutionarily and functionally related P450 enzyme 17-hydroxylase, which has a large range of substrates.

Earlier studies of the human enzyme expressed in yeast classified 17-hydroxyprogesterone as the best substrate for 21-hydroxylase. However, recent analysis of the purified human enzyme found a lower K_M and greater catalytic efficiency for progesterone over 17-hydroxyprogesterone.

The 2015 analysis found the catalytic efficiency of 21-hydroxylase for conversion of progesterone in humans to be approximately 1.3 x 10^7 M-1s-1 at 37 °C. This makes it the most catalytically efficient P450 enzyme of those reported, as of 2015, and more catalytically efficient than the closely related bovine 21-hydroxylase enzyme. C-H bond breaking to create a primary carbon radical is thought to be the rate-limiting step in the hydroxylation.

Pathway

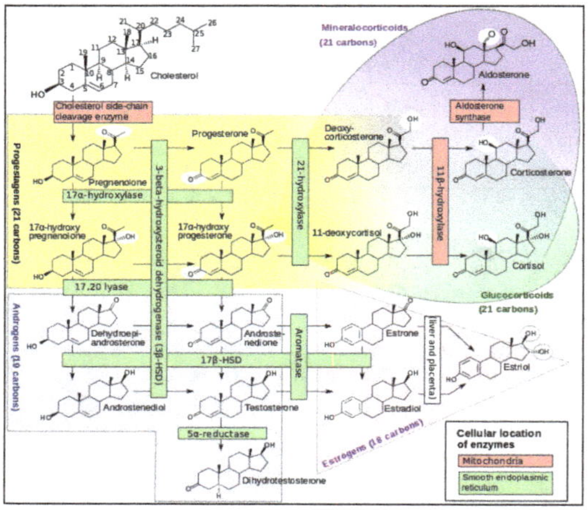

Human steroidogenesis, showing both reactions of 21-Hydroxylase at center top.

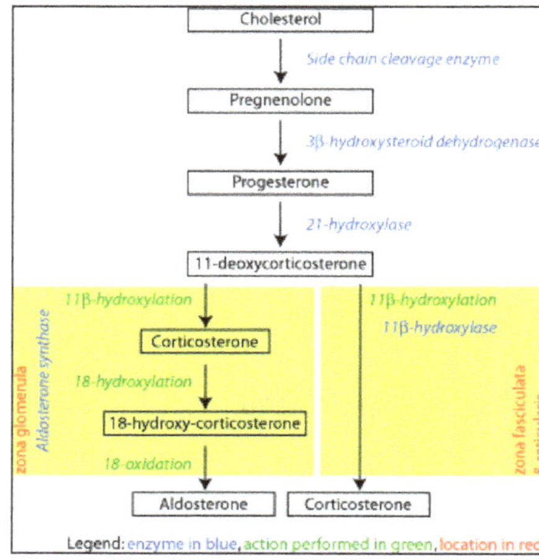

Corticosteroid biosynthetic pathway in the rat.

Clinical Significance

A defect within the CYP21A2 gene causes a disturbance of the development of the enzyme, which leads to congenital adrenal hyperplasia due to 21-hydroxylase deficiency. A related pseudogene is located near this gene; gene conversion events involving the functional gene and the pseudogene are thought to account for many cases of steroid 21-hydroxylase deficiency. Both genes are located on chromosome 6, in the major histocompatibility complex, and the pseudogene, CYP21A1, retains 98% exonic sequence identity with the functional gene.

Congenital adrenal hyperplasia (CAH) is an autosomal recessive disorder, and occurs in approximately 1 in 15000 births globally. There are multiple forms of CAH, broken down into classical and nonclassical forms based on the amount of function retained. The classical forms include salt-wasting (SW), and simple-viralizing (SV). Mutations that interfere with the active site—the heme group or residues involved in substrate binding—result in a complete loss of enzymatic activity, the salt-wasting type. Cortisol and aldosterone deficits are associated with life-threatening salt-loss (hence salt-wasting), as the steroids play roles in regulating sodium homeostasis. Retaining minimal enzyme activity, the simple-viralizing type is associated with mutations in conserved hydrophobic regions or near the transmembrane domain. Simple viralizing CAH patients maintain adequate sodium homeostasis, but exhibit other phenotypical symptoms shared by SW, including accelerated growth in childhood and ambiguous genitalia in female neonates. Nonclassical forms retain 20-60% of hydroxylase function—this form is associated with normal cortisol expression, but an excess of androgens post-puberty.

ADENOSINE DEAMINASE

Adenosine deaminase (also known as adenosine aminohydrolase, or ADA) is an enzyme involved in purine metabolism. It is needed for the breakdown of adenosine from food and for the turnover of nucleic acids in tissues.

Its primary function in humans is the development and maintenance of the immune system. However, the full physiological role of ADA is not yet completely understood.

Structure

ADA exists in both small form (as a monomer) and large form (as a dimer-complex). In the monomer form, the enzyme is a polypeptide chain, folded into eight strands of parallel α/β barrels, which surround a central deep pocket that is the active site. In addition to the eight central β-barrels and eight peripheral α-helices, ADA also contains five additional helices: residues 19-76 fold into three helices, located between β1 and α1 folds; and two antiparallel carboxy-terminal helices are located across the amino-terminal of the β-barrel.

The ADA active site contains a zinc ion, which is located in the deepest recess of the active site and coordinated by five atoms from His15, His17, His214, Asp295, and the substrate. Zinc is the only cofactor necessary for activity.

The substrate, adenosine, is stabilized and bound to the active site by nine hydrogen bonds. The carboxyl group of Glu217, roughly coplanar with the substrate purine ring, is in position to form a hydrogen bond with N1 of the substrate. The carboxyl group of Asp296, also coplanar with the substrate purine ring, forms hydrogen bond with N7 of the substrate. The NH group of Gly184 is in position to form a hydrogen bond with N3 of the substrate. Asp296 forms bonds both with the Zn^{2+} ion as well as with 6-OH of the substrate. His238 also hydrogen bonds to substrate 6-OH. The 3′-OH of the substrate ribose forms a hydrogen bond with Asp19, while the 5′-OH forms a hydrogen bond with His17. Two further hydrogen bonds are formed to water molecules, at the opening of the active site, by the 2′-OH and 3′-OH of the substrate.

Due to the recessing of the active site inside the enzyme, the substrate, once bound, is almost completely sequestered from solvent. The surface exposure of the substrate to solvent when bound is 0.5% the surface exposure of the substrate in the free state.

Reactions

ADA irreversibly deaminates adenosine, converting it to the related nucleoside inosine by the substitution of the amino group for a keto group.

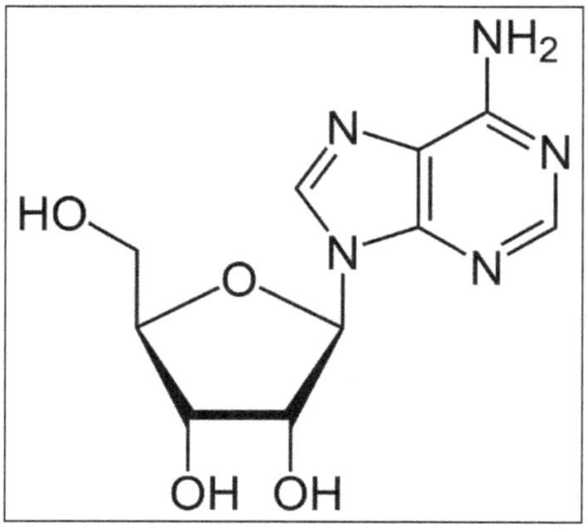

Adenosine.

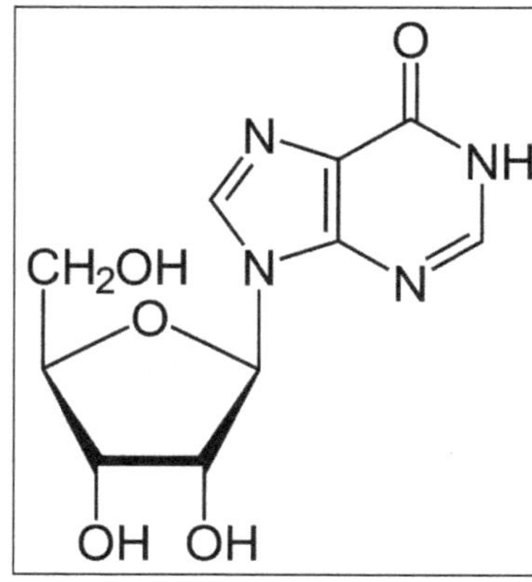

Inosine.

Inosine can then be deribosylated (removed from ribose) by another enzyme called purine nucleoside phosphorylase (PNP), converting it to hypoxanthine.

Mechanism of Catalysis

The proposed mechanism for ADA-catalyzed deamination is stereospecific addition-elimination via tetrahedral intermediate. By either mechanism, Zn^{2+} as a strong electrophile activates a water molecule, which is deprotonated by the basic Asp295 to form the attacking hydroxide. His238 orients the water molecule and stabilizes the charge of the attacking hydroxide. Glu217 is protonated to donate a proton to N1 of the substrate.

The reaction is stereospecific due to the location of the zinc, Asp295, and His238 residues, which all face the B-side of the purine ring of the substrate.

Competitive inhibition has been observed for ADA, where the product inosine acts at the competitive inhibitor to enzymatic activity.

Function

ADA is considered one of the key enzymes of purine metabolism. The enzyme has been found in bacteria, plants, invertebrates, vertebrates, and mammals, with high conservation of amino acid sequence. The high degree of amino acid sequence conservation suggests the crucial nature of ADA in the purine salvage pathway.

Primarily, ADA in humans is involved in the development and maintenance of the immune system. However, ADA association has also been observed with epithelial cell differentiation, neurotransmission, and gestation maintenance. It has also been proposed that ADA, in addition to adenosine breakdown, stimulates release of excitatory amino acids and is necessary to the coupling of A1 adenosine receptors and heterotrimeric G proteins. Adenosine deaminase deficiency leads to pulmonary fibrosis, suggesting that chronic exposure to high levels of adenosine can exacerbate inflammation responses rather than suppressing them. It has also been recognized that adenosine deaminase protein and activity is upregulated in mouse hearts that overexpress HIF-1 alpha, which in part explains the attenuated levels of adenosine in HIF-1 alpha expressing hearts during ischemic stress.

Pathology

Some mutations in the gene for adenosine deaminase cause it not to be expressed. The resulting deficiency is one cause of severe combined immunodeficiency (SCID), particularly of autosomal recessive inheritance. Deficient levels of ADA have also been associated with pulmonary inflammation, thymic cell death, and defective T-cell receptor signaling.

Conversely, mutations causing this enzyme to be overexpressed are one cause of hemolytic anemia.

There is some evidence that a different allele (ADA2) may lead to autism. Elevated levels of ADA has also been associated with AIDS.

Isoforms

There are 2 isoforms of ADA: ADA1 and ADA2.

- ADA1 is found in most body cells, particularly lymphocytes and macrophages, where it is present not only in the cytosol and nucleus but also as the ecto- form on the cell membrane attached to dipeptidyl peptidase-4 (aka, CD26). ADA1 is involved mostly in intracellular activity, and exists both in small form (monomer) and large form (dimer). The interconversion of small to large forms is regulated by a 'conversion factor' in the lung.
- ADA2 was first identified in human spleen. It was subsequently found in other tissues including the macrophage where it co-exists with ADA1. The two isoforms regulate the ratio of adenosine to deoxyadenosine potentiating the killing of parasites. ADA2 is found predominantly in the human plasma and serum, and exists solely as a homodimer.

Clinical Significance

ADA2 is the predominant form present in human blood plasma and is increased in many diseases, particularly those associated with the immune system: for example rheumatoid arthritis, psoriasis, and sarcoidosis. The plasma ADA2 isoform is also increased in most cancers. ADA2 is not ubiquitous but co-exists with ADA1 only in monocytes-macrophages.

Total plasma ADA can be measured using high performance liquid chromatography or enzymatic or colorimetric techniques. Perhaps the simplest system is the measurement of the ammonia released from adenosine when broken down to inosine. After incubation of plasma with a buffered solution of adenosine the ammonia is reacted with a Berthelot reagent to form a blue colour which is proportionate to the amount of enzyme activity. To measure ADA2, erythro-9-(2-hydroxy-3-nonyl) adenine (EHNA) is added prior to incubation so as to inhibit the enzymatic activity of ADA1. It is the absence of ADA1 that causes SCID.

ADA can also be used in the workup of lymphocytic pleural effusions or peritoneal ascites, in that such specimens with low ADA levels essentially excludes tuberculosis from consideration.

Tuberculosis pleural effusions can now be diagnosed accurately by increased levels of pleural fluid adenosine deaminase, above 40 U per liter.

Cladribine and Pentostatin are anti-neoplastic agents used in the treatment of hairy cell leukemia; their mechanism of action is inhibition of adenosine deaminase.

ADRENODOXIN REDUCTASE

Adrenodoxin reductase (Enzyme Nomenclature name: adrenodoxin-NADP+ reductase), was first isolated from bovine adrenal cortex where it functions as the first enzyme in the mitochondrial P450 systems that catalyze essential steps in steroid hormone biosynthesis. Examination of complete genome sequences revealed that adrenodoxin reductase gene is present in most metazoans and prokaryotes.

Nomenclature

The name of the enzyme was coined based on its function to reduce a [2Fe-2S] (2 iron, 2 sulfur) electron-transfer protein that was named adrenodoxin. Later, in some studies, the enzyme was also referred to as a "ferredoxin reductase", as adrenodoxin is a ferredoxin. In the human gene nomenclature, the standard name is ferredoxin reductase and the symbol is FDXR, with ADXR specified as a synonym.

The assignment of the name "ferredoxin reductase" has been criticized as a misnomer because determination of the structure of adrenodoxin reductase revealed that it is completely different from that of plant ferredoxin reductase and there is no homology between these two enzymes. With more proteins with a ferroxodin-reducing activity discovered in both families as well as novel families, this enzyme activity is now seen as an example of convergent evolution.

Function

Adrenodoxin reductase is a flavoprotein as it carries a FAD type coenzyme. The enzyme functions as the first electron transfer protein of mitochondrial P450 systems such as

P450scc. The FAD coenzyme receives two electrons from NADPH and transfers them one at a time to the electron transfer protein adrenodoxin. Adrenodoxin functions as a mobile shuttle that transfers electrons between ADXR and mitochondrial P450s.

It catalyzes the following reaction:

$$\text{NADPH} + 2\ \text{oxidized adrenodoxin} \longrightarrow 2\ \text{reduced adrenodoxin} + \text{NADP}^+ + \text{H}^+$$

Gene Structure

The cDNA for adrenodoxin reductase was first cloned in 1987. In both bovine and human genomes there is only a single copy of the gene.

Sites of Expression

ADXR gene is expressed in all tissues that have mitochondrial P450s. The highest levels of the enzyme are found in the adrenal cortex, granulosa cells of the ovary and leydig cells of the testis that specialize in steroid hormone synthesis. Immmunofluorescent staining shows that enzyme is localized in mitochondria. The enzyme is also expressed in the liver, the kidney and the placenta.

Enzyme Structure

Adrenodoxin reductase has two domains that bind NADPH and FAD separately. The FAD and NADP binding sites of the enzyme were predicted by sequence analysis of the enzyme.

While the FAD-binding site has a consensus sequence (Gly-x-Gly-x-x-Gly) that is similar to other Rossmann folds in FAD and NAD binding sites, the NADPH binding site consensus sequence differs from the FAD-binding site by the substitution of an alanine instead of the last Gly (Gly-x-Gly-x-x-Ala). The location of these FAD and NADP binding sites were confirmed by the crystal structure of the enzyme.

ALDEHYDE DEHYDROGENASE

Aldehyde dehydrogenases are a group of enzymes that catalyse the oxidation of aldehydes. They convert aldehydes (R–C(=O)–H) to carboxylic acids (R–C(=O)–O–H). The oxygen comes from a water molecule. To date, nineteen ALDH genes have been identified within the human genome. These genes participate in a wide variety of biological processes including the detoxification of exogenously and endogenously generated aldehydes.

Function

Aldehyde dehydrogenase is a polymorphic enzyme responsible for the oxidation of aldehydes to carboxylic acids, which leave the liver and are metabolized by the body's muscle and heart. There are three different classes of these enzymes in mammals: class 1 (low K_m, cytosolic), class 2 (low K_m, mitochondrial), and class 3 (high K_m, such as those expressed in tumors, stomach, and cornea). In all three classes, constitutive and inducible forms exist. ALDH1 and ALDH2 are the most important enzymes for aldehyde oxidation, and both are tetrameric enzymes composed of 54 kDa subunits. These enzymes are found in many tissues of the body but are at the highest concentration in the liver.

Active Site

The active site of the aldehyde dehydrogenase enzyme is largely conserved throughout the different classes of the enzyme and, although the number of amino acids present in a subunit can change, the overall function of the site changes little. The active site binds to one molecule of an aldehyde and one of either NAD^+ or $NADP^+$ that functions as a cofactor. A cysteine and a glutamate will interact with the aldehyde substrate. Many other residues will interact with the $NAD(P)^+$ to hold it in place. A magnesium may be used to help the enzyme function, although the amount it helps the enzyme can vary between different classes of aldehydes.

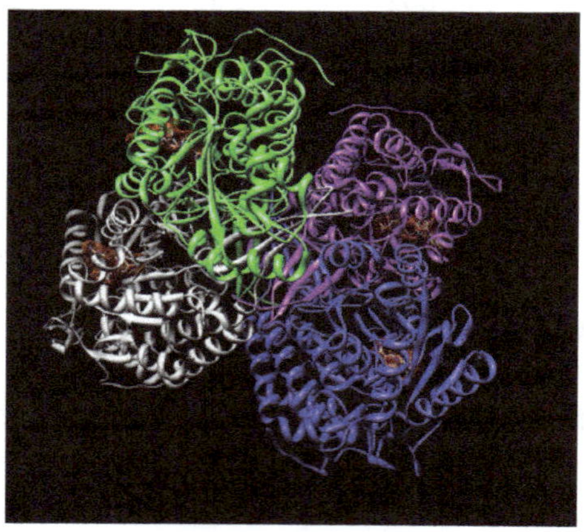

Tetramer of aldehyde dehydrogenase 2 with a space filling model of NAD^+ in each active site.

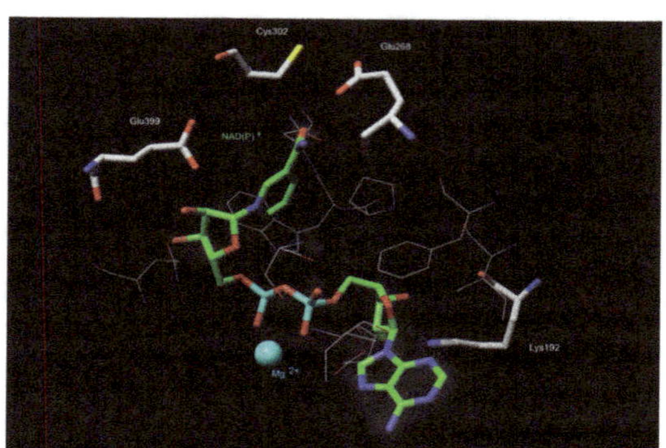

The active site of a human mitochondrial aldehyde dehydrogenase 2. Cys302 and Glu268 interact with the aldehyde substrate. The NAD^+ is held in place by multiple residues (shown as wires or sticks).

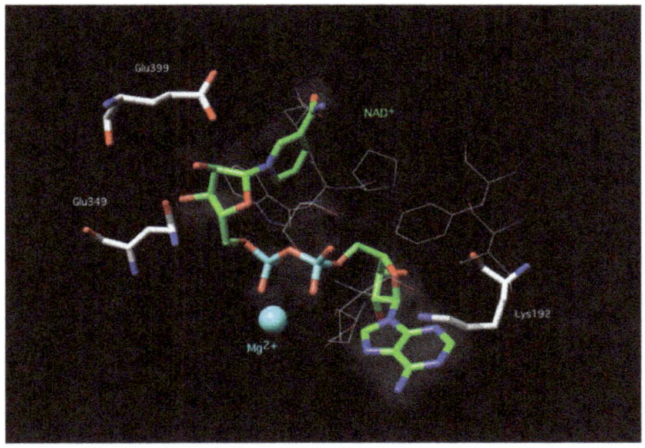

The active site of the K487E mutant aldehyde dehydrogenase 2 with a space filling model of NAD+ in the active site. The amino acid Glu349 is highlighted.

Mechanism

The overall reaction catalysed by the aldehyde dehydrogenases is:

- $RCHO + NAD^+ + H_2O \rightarrow RCOOH + NADH + H^+$

In this NAD(P)+-dependent reaction, the aldehyde enters the active site through a channel extending from the surface of the enzyme. The active site contains a Rossman fold, and interactions between the cofactor and the fold allow for the action of the active site.

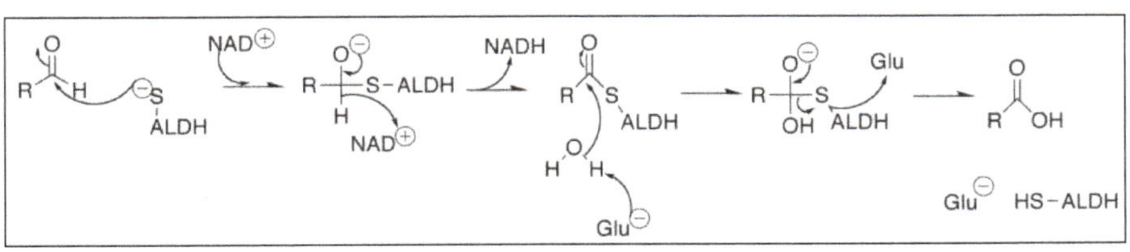

A sulfur from a cysteine in the active site makes a nucleophilic attack on the carbonyl carbon of the aldehyde. The hydrogen is kicked off as a hydride and attacks NAD(P)+ to make NAD(P)H. The enzyme's active site then goes through an isomorphic change whereby the NAD(P)H is moved, creating room for a water molecule to access the substrate. The water is primed by a glutamate in the active site, and the water makes a nucleophilic attack on the carbonyl carbon, kicking off the sulfur as a leaving group.

Pathology (Aldehyde Dehydrogenase Deficiency)

Role of aldehyde dehydrogenase (shown in red box) in norepinephrine degradation, creating vanillylmandelic acid, the major catecholamine metabolite.

ALDH2 plays a crucial role in maintaining low blood levels of acetaldehyde during alcohol oxidation. In this pathway, the intermediate structures can be toxic, and health problems arise when those intermediates cannot be cleared. When high levels of acetaldehyde occur in the blood, facial flushing, lightheadedness, palpitations, nausea, and general "hangover" symptoms occur. These symptoms are indicative of a medical condition known as the alcohol flush reaction, also known as "Asian flush" or "Oriental flushing syndrome".

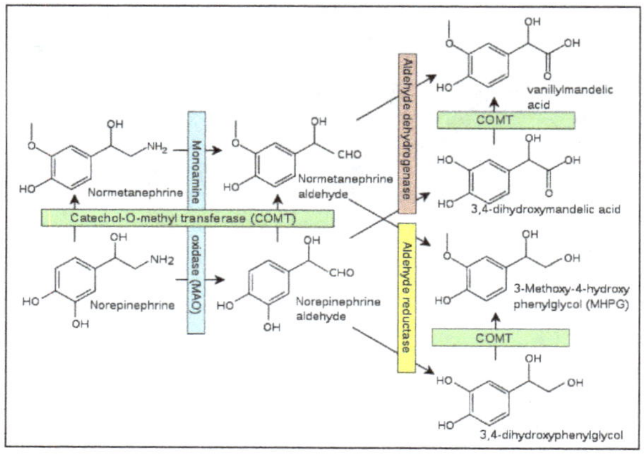

There is a mutant form of aldehyde dehydrogenase, termed ALDH2*2, wherein a lysine residue replaces a glutamate in the active site at position 487 of ALDH2. Homozygous individuals with the mutant allele have almost no ALDH2 activity, and those heterozygous for the mutation have reduced activity. Thus, the mutation is partially dominant. The ineffective homozygous allele works at a rate of about 8% of the normal allele, for it shows a higher K_m for NAD^+ and has a higher maximum velocity than the wild-type allele. This mutation is common in Japan, where 41% of a non-alcoholic control group were ALDH2 deficient, where only 2–5% of an alcoholic group were ALDH2-deficient. In Taiwan, the numbers are similar, with 30% of the control group showing the deficiency and 6% of alcoholics displaying it. The deficiency is manifested by slow acetaldehyde removal, with low alcohol tolerance perhaps leading to a lower frequency of alcoholism.

These symptoms are the same as those observed in people who drink while being treated by the drug disulfiram, which is why it is used to treat alcoholism. The patients show higher blood levels of acetaldehyde, and become violently ill upon consumption of even small amounts of alcohol. Several drugs (e.g., metronidazole) cause a similar reaction known as "disulfiram-like reaction."

Yokoyama *et al.* found that decreased enzyme activity of aldehyde dehydrogenase-2, caused by the mutated ALDH2 allele, contributes to a higher chance of esophageal and oropharyngolaryngeal cancers. The metabolized acetaldehyde in the blood, which is six times higher than in individuals without the mutation, has shown to be a carcinogen in lab animals. ALDH2*2 is associated with increased odds of oropharyngolaryngeal, esophageal, gastric, colon, and lung cancer. However, they found no connection between increased levels of ALDH2*2 in the blood and an increased risk of liver cancer.

Some case-control studies claimed that carriage of ALDH2*2 allele was a risk of late-onset Alzheimer's disease independent of the apolipoprotein E gene (the odds for LOAD in carriers of ALDH2*2 allele almost twice that of non-carriers). Moreover, ALDH gene, protein expression and activity are substantially decreased in the substantia nigra of Parkinson's disease patients. These reports are in line with findings implementing toxic lipid oxidation-derived aldehydes in these diseases and in neurodegeneration in general.

Fitzmaurice *et al.* explored aldehyde dehydrogenase inhibition as a pathogenic mechanism in Parkinson disease. "This ALDH model for PD etiology may help explain the selective vulnerability of dopaminergic neurons in PD and provide a potential mechanism through which environmental toxicants contribute to PD pathogenesis."

Knockout mouse models further confirm the involvement of ALDH family in neurodegeneration. Mice null for ALDH1a1 and ALDH2 exhibit Parkinson's disease-like age-dependent deficits in motor performance and significant increase in biogenic aldehydes.

The ALDH2-/- mice display age-related memory deficits in various tasks, as well as endothelial dysfunction, brain atrophy, and other Alzheimer's disease-associated pathologies, including marked increases in lipid peroxidation products, amyloid-beta, p-tau and activated caspases. These behavioral and biochemical Alzheimer's disease-like deficits were efficiently ameliorated when the ALDH2-/- mice were treated with isotope-reinforced, deuterated polyunsaturated fatty acids (D-PUFA).

ALPHA-AMYLASE

Alpha-amylase, (α-amylase) is an enzyme that hydrolyses alpha bonds of large, alpha-linked polysaccharides, such as starch and glycogen, yielding glucose and maltose. It is the major form of amylase found in humans and other mammals. It is also present in seeds containing starch as a food reserve, and is secreted by many fungi.

Although found in many tissues, amylase is most prominent in pancreatic juice and saliva, each of which has its own isoform of human α-amylase. They behave differently on isoelectric focusing, and can also be separated in testing by using specific monoclonal antibodies. In humans, all amylase isoforms link to chromosome 1p21.

Salivary Amylase (Ptyalin)

Amylase is found in saliva and breaks starch into maltose and dextrin. This form of amylase is also called "ptyalin", which was named by Swedish chemist Jöns Jacob Berzelius. It will break large, insoluble starch molecules into soluble starches (amylodextrin, erythrodextrin, and achrodextrin) producing successively smaller starches and ultimately maltose. Ptyalin acts on linear α(1,4) glycosidic linkages, but compound hydrolysis requires an enzyme that acts on branched products. Salivary amylase is inactivated in the stomach by gastric acid. In gastric juice adjusted to pH 3.3, ptyalin was totally inactivated in 20 minutes at 37 °C. In contrast, 50% of amylase activity remained after 150 minutes of exposure to gastric juice at pH 4.3. Both starch, the substrate for ptyalin, and the product (short chains of glucose) are able to partially protect it against inactivation by gastric acid. Ptyalin added to buffer at pH 3.0 underwent complete inactivation in 120 minutes; however, addition of starch at a 0.1% level resulted in 10% of the activity remaining, and similar addition of starch to a 1.0% level resulted in about 40% of the activity remaining at 120 minutes.

Optimum Conditions for Ptyalin

- Optimum pH – 7.0.
- Human body temperature.
- Presence of certain anions and activators.
- Chloride and bromide – most effective.
- Iodide – less effective.
- Sulfate and phosphate – least effective.

Genetic Variation in Human Salivary Amylase

The salivary amylase gene has undergone duplication during evolution, and DNA hybridization studies indicate many individuals have multiple tandem repeats of the gene. The number of gene copies correlates with the levels of salivary amylase, as measured by protein blot assays using antibodies to human amylase. Gene copy number is associated with apparent evolutionary exposure to high-starch diets. For example, a Japanese individual had 14 copies of the amylase gene (one allele with 10 copies, and a second allele with four copies). The Japanese diet has traditionally contained large amounts of rice starch. In contrast, a Biaka individual carried six copies (three copies on each allele). The Biaka are rainforest hunter-gatherers who have traditionally consumed a low-starch diet. Perry and colleagues speculated the increased copy number of the salivary amylase gene may have enhanced survival coincident to a shift to a starchy diet during human evolution.

Pancreatic Amylase

Pancreatic α-amylase randomly cleaves the α(1-4) glycosidic linkages of amylose to yield dextrin, maltose, or maltotriose. It adopts a double displacement mechanism with retention of anomeric configuration.

In Pathology

The test for amylase is easier to perform than that for lipase, making it the primary test used to detect and monitor pancreatitis. Medical laboratories will usually measure either pancreatic amylase or total amylase. If only pancreatic amylase is measured, an increase will not be noted with mumps or other salivary gland trauma.

However, because of the small amount present, timing is critical when sampling blood for this measurement. Blood should be taken soon after a bout of pancreatitis pain, otherwise it is excreted rapidly by the kidneys.

Salivary α-amylase has been used as a biomarker for stress and as a surrogate marker of sympathetic nervous system (SNS) activity that does not require a blood draw.

Interpretation

Increased plasma levels in humans are found in:

- Salivary trauma (including anaesthetic intubation).
- Mumps – due to inflammation of the salivary glands.
- Pancreatitis – because of damage to the cells that produce amylase.
- Renal failure – due to reduced excretion.

Total amylase readings of over 10 times the upper limit of normal (ULN) are suggestive of pancreatitis. Five to 10 times the ULN may indicate ileus or duodenal disease or renal failure, and lower elevations are commonly found in salivary gland disease.

Genes

- Salivary – *AMY1A, AMY1B, AMY1C*
- Pancreatic – *AMY2A, AMY2B*

In Grain

α-Amylase activity in grain is measured by, for instance, the Hagberg-Perten Falling Number, a test to assess sprout damages, or the Phadebas method.

Industrial Use

α-Amylase is used in ethanol production to break starches in grains into fermentable sugars.

The first step in the production of high-fructose corn syrup is the treatment of cornstarch with α-amylase, producing shorter chains of sugars oligosaccharides.

An α-amylase called "Termamyl", sourced from *Bacillus licheniformis,* is also used in some detergents, especially dishwashing and starch-removing detergents.

Potential for Medical Use

α-Amylase has exhibited efficacy in degrading polymicrobial bacterial biofilms by hydrolyzing the α(1-4) glycosidic linkages within the structural, matrix exopolysaccharides of the extracellular polymeric substance (EPS).

Buffer Inhibition

The tris molecule is reported to inhibit a number of bacterial α-amylases, so they should not be used in tris buffer.

Determination

Several methods are available for determination of α-amylase activity, and different industries tend to rely on different methods. The starch iodine test, a development of the iodine test, is based on colour change, as α-amylase degrades starch and is commonly used in many applications. A similar but industrially produced test is the Phadebas amylase test, which is used as a qualitative and quantitative test within many industries, such as detergents, various flour, grain, and malt foods, and forensic biology.

Domain Architecture

α-Amylases contain a number of distinct protein domains. The catalytic domain has a structure consisting of an eight-stranded alpha/beta barrel that contains the active site, interrupted by a ~70-amino acid calcium-binding domain protruding between beta strand 3 and alpha helix 3, and a carboxyl-terminal Greek key beta-barrel domain. Several alpha-amylases contain a beta-sheet domain, usually at the C terminus. This domain is organised as a five-stranded antiparallel beta-sheet. Several alpha-amylases contain an all-beta domain, usually at the C terminus.

CAPPING ENZYME

A capping enzyme (CE) is an enzyme that catalyzes the attachment of the 5′ cap to messenger RNA molecules that are in the process of being synthesized in the cell nucleus during the first stages of gene expression. The addition of the cap occurs co-transcriptionally, after the growing RNA molecule contains as little as 25 nucleotides. The enzymatic reaction is catalyzed specifically by the phosphorylated carboxyl-terminal domain (CTD) of RNA polymerase II. The 5′ cap is therefore specific to RNAs synthesized by this polymerase rather than those synthesized by RNA polymerase I or RNA polymerase III. Pre-mRNA undergoes a series of modifications - 5′ capping, splicing and 3′ polyadenylation before becoming mature mRNA that exits the nucleus to be translated into functional proteins and capping of the 5′ end is the first of these modifications. Three enzymes, RNA triphosphatase, guanylyltransferase (or CE), and methyltransferase are involved in the addition of the methylated 5′ cap to the mRNA.

Formation of the Cap

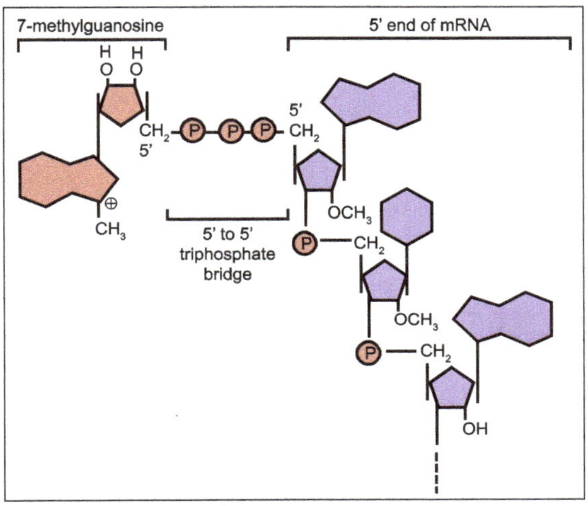

5′ cap structure.

Capping is a three-step process that utilizes the enzymes RNA triphosphatase, guanylyltransferase, and methyltransferase. Through a series of three steps, the cap is added to the first nucleotide's 5′ hydroxyl group of the growing mRNA strand while transcription is still occurring. First, RNA 5′ triphosphatase hydrolyzes the 5′ triphosphate group to make diphospate-RNA. Then, the addition of GMP by guanylyltransferase produces the guanosine cap. Last, RNA methyltransferase transfers a methyl group to the guanosine cap to yield 7-methylguanosine cap that is attached to the 5′ end of the transcript. These three enzymes, collectively called the capping enzymes, are only able to catalyze their respective reactions when attached to RNA polymerase II, an enzyme necessary for the transcription of DNA into pre-mRNA. When this complex of RNA polymerase II and the capping enzymes is achieved, the capping enzymes are able to add the cap to the mRNA while it is produced by RNA polymerase II.

Function

Eukaryotic RNA must undergo a series of modifications in order to be exported from the nucleus and successfully translated into function proteins, many of which are dependent on mRNA capping, the first mRNA modification to take place. 5′ capping is essential for mRNA stability, enhancing mRNA processing, mRNA export and translation. After successful capping, an additional phosphorylation event initiates the recruitment of machinery necessary for RNA splicing, a process by which introns are removed to produce a mature mRNA. The addition of the cap onto mRNA confers protection to the transcript from exonucleases that degrade unprotected RNA and assist in the nuclear export transport process so that the mRNA can be translated to form proteins. The function of the 5′ cap is essential to the ultimate expression of the RNA.

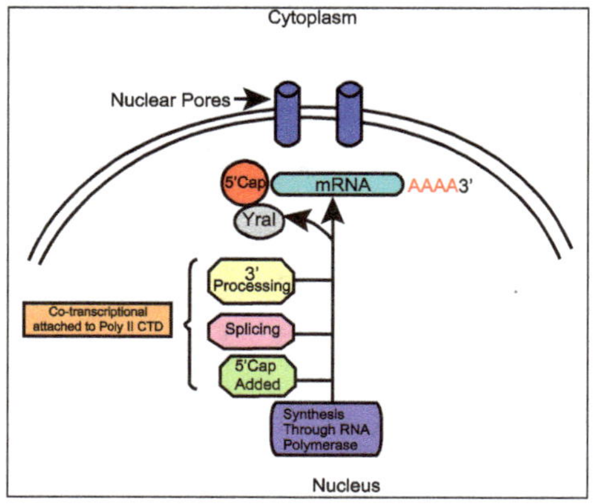

An illustration of how mRNA is processed for export, starting with the 5′capping process.

Structure

The capping enzyme is part of the covalent nucleotidyl transferases superfamily, which also includes DNA ligases and RNA ligases. The enzymes of this superfamily share the following similarities:

- Conserved regions known as motifs I, II, III, IIIa, IV, V and VI, which are arranged in the same order and similar spacing.
- A lysine containing motif KxDG (motif I).
- A covalent lysyl-NMP intermediate.

The capping enzyme is composed of two domains, a nucleotidyl transferase (NTase) domain and a C-terminal oligonucleotide binding (OB) domain. The NTase domain, conserved in capping enzymes, DNA and RNA ligases, is made up 5 motifs, I, III, IIIa, IV and V. Motif I or KxDG is the active site where the covalent (lysyl)-N-GMP intermediate is formed. Both the NTase and OB domains undergo conformational changes that assist in the capping reaction.

Capping enzymes are found in the nucleus of eukaryotic cells. Depending on the organism, the capping enzyme is either a monofunctional or bifunctional polypeptide. The guanylyltransferases (Ceg1) of *Saccharomyces cerevisiae* is encoded by the *CEG1* gene and is composed of 459 amino acids (53-kD). The RNA triphosphatase (Cet1) is a separate 549 amino acid polypeptide (80-kD), encoded by the *CET1* gene. The human capping enzyme is an example of a bifunctional polypeptide, which has both triphosphatase (N-terminal) and guanylyltransferase (C-teriminal) domains. The human mRNA guanylyltransferase domain of the capping enzyme is composed of seven helices and fifteen β strands that are grouped into three, five and seven strands, arranged as antiparallel β sheets. The enzyme structure has three sub-domains referred to hinge, base and lid. The GTP binding site is located between the hinge and base domain. The lid domain determines the conformation of the active site cleft, which consists of the GTP binding site, phosphoamide linking lysine and surrounding residues. The guanylyltransferase domain is linked to the triphosphatase domain via a 25 amino acid flexible loop structure.

Impact of the Enzyme's Activity

Splicing is dependent on the presence of the 7-methylguanosine cap. A defect in splicing can occur as a result of mutation(s) in the guanylytransferase, which can inhibit enzyme activity, preventing the formation of the cap. However the severity of the effect is dependent on the guanlyltransferase mutation. Furthermore, the guanylyltransferase relieves transcriptional repression mediated by NELF. NELF together with DSIF prevents transcription elongation. Thus, mutations in the enzyme can affect transcription elongation.

LIPOXYGENASE

Lipoxygenases are a family of (non-heme) iron-containing enzymes most of which catalyze the dioxygenation of polyunsaturated fatty acids in lipids containing a cis,cis-1,4-pentadiene into cell signaling agents that serve diverse roles as autocrine signals that regulate the function of their parent cells, paracrine signals that regulate the function of nearby cells, and endocrine signals that regulate the function of distant cells.

The lipoxygenases are related to each other based upon their similar genetic structure and dioxygenation activity. However, one lipoxygenase, ALOXE3, while having a lipoxygenase genetic structure, possesses relatively little dioxygenation activity; rather its primary activity appears to be as an isomerase that catalyzes the conversion of hydroperoxy unsaturated fatty acids to their 1,5-epoxide, hydroxyl derivatives.

Lipoxygenases are found in eukaryotes (plants, fungi, animals, protists); while the third domain of terrestrial life, the archaea, possesses proteins with a slight (~20%) amino acid sequence similarity to lipoxygenases, these proteins lack iron-binding residues and therefore are not projected to possess lipoxygenase activity.

Biochemistry

Based on detailed analyses of 15-lipoxygenase 1 and stabilized 5-lipoxygenase, lipoxygenase structures consist of a 15 kilodalton N-terminal beta barrel domain, a small (e.g. ~0.6 kilodalton) linker inter-domain, and a relatively large C-terminal catalytic domain which contains the non-heme iron critical for the enzymes' catalytic activity. Most of the lipoxygenases (exception, ALOXE3) catalyze the reaction *Polyunsaturated fatty acid* + O_2 → *fatty acid hydroperoxide* in four steps:

- The rate-limiting step of hydrogen abstraction from a bisallylic methylene carbon to form a fatty acid radical at that carbon.
- Rearrangement of the radical to another carbon center.
- Addition of molecular oxygen (O_2) to the rearranged carbon radical center thereby forming a peroxy radical(—OO·) bond to that carbon.
- Reduction of the peroxy radical to its corresponding anion ($—OO^-$).

The ($—OO^-$) residue may then be protonated to form a hydroperoxide group (—OOH) and further metabolized by the lipoxygenase to e.g. leukotrienes, hepoxilins, and various specialized pro-resolving mediators, or reduced by ubiquitous cellular glutathione peroxidases to a hydroxy group thereby forming hydroxylated (—OH) polyunsaturated fatty acids such as the Hydroxyeicosatetraenoic acids and HODEs (i.e. hydroxyoctadecaenoic acids).

Polyunsaturated fatty acids that serve as substrates for one or more of the lipoxygenases include the omega 6 fatty acids, arachidonic acid, linoleic acid, dihomo-γ-linolenic acid, and adrenic acid; the omega-3 fatty acids, eicosapentaenoic acid, docosahexaenoic acid, and alpha-linolenic acid; and the omega-9 fatty acid, mead acid. Certain types of the lipoxygenases, e.g. human and murine 15-lipoxygenase 1, 12-lipoxygenase B, and ALOXE3, are capable of metabolizing fatty acid substrates that are constituents of phospholipids, cholesterol esters, or complex lipids of the skin. Most lipoxygenases catalyze the formation of initially formed hydroperoxy products that have *S* chirality. Exceptions to this rule include the 12R-lipoxygenases of humans and other mammals.

Lipoxygenases depend on the availability of their polyunsaturated fatty acid substrates which, particularly in mammalian cells, is normally maintained at extremely low levels. In general, various phospholipase A2s and diacylglycerol lipases are activated during cell stimulation, proceed to release these fatty acids from their storage sites, and thereby are key regulators in the formation of lipoxygenase-dependent metabolites. In addition, cells, when so activated, may transfer their released polyunsaturated fatty acids to adjacent or nearby cells which then metabolize them through their lipoxygenase pathways in a process termed transcellular metabolism or transcellular biosynthesis.

Biological Function and Classification

These enzymes are most common in plants where they may be involved in a number of diverse aspects of plant physiology including growth and development, pest resistance, and senescence or responses to wounding. In mammals a number of lipoxygenases isozymes are involved in the metabolism of eicosanoids (such as prostaglandins, leukotrienes and nonclassic eicosanoids). Sequence data is available for the following lipoxygenases:

Plant Lipoxygenases

Plants express a variety of cytosolic lipoxygenases as well as what seems to be a chloroplast isozyme. Plant lipoxygenase in conjunction with hydroperoxide lyases are responsible for many fragrances and other signalling compounds. One example is cis-3-hexenal, the odor of freshly cut grass.

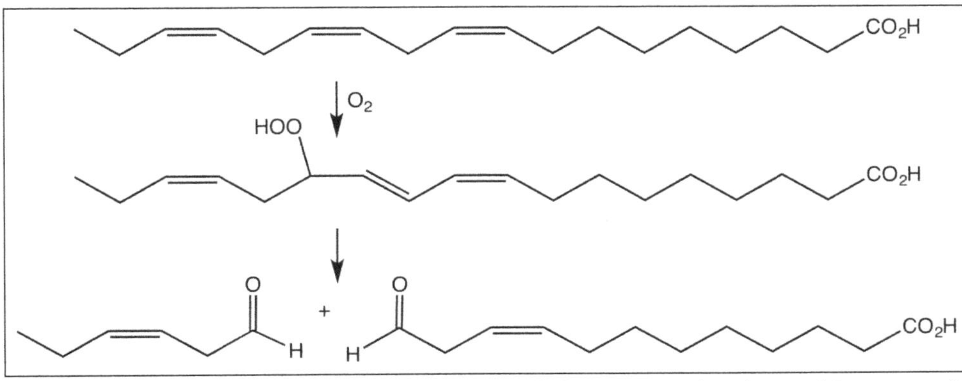

An illustrative transformation involving a hydroperoxide lyase. Here cis-3-hexenal is generated from linolenic acid to the hydroperoxide by the action of a lipoxygenase followed by the lyase.

Human Lipoxygenases

With the exception of the 5-LOX gene which is located on chromosome 10q11.2, all six human LOX genes are located on chromosome 17.p13 and code for a single chain protein of 75–81 kiloDaltons and consisting of 662–711 amino acids. Mammalian LOX genes contain 14 (ALOX5, ALOX12, ALOX15, ALOX15B) or 15 (ALOX12B, ALOXE3) exons with exon/intron boundaries at highly conserved position. The 6 human lipoxygenases along with some of the major products that they make as well as some their associations with genetic diseases are as follows:

- Arachidonate 5-lipoxygenase (ALOX5), also termed 5-lipoxygenase, 5-LOX, and 5-LO. Major products: it metabolizes arachidonic acid to 5-hydroperoxy-eicostetraeoic acid (5-HpETE) which is converted to 1) 5-Hydroxyicosatetraenoic acid (5-HETE) and then to 5-oxo-eicosatetraenoic acid (5-oxo-ETE), 2) leukotriene A4 (LTBA) which may then be converted to leukotriene B4 (LTB4) or Leukotriene C4 (LTC4) (LTC4 may be further metabolized to leukotriene D4 [LTD4] and then to Leukotriene E4 [LTE4]), or 3 acting in series with ALOX15, to the Specialized pro-resolving mediators, lipoxins A4 and B4. ALOX5 also metabolizes eicosapentaenoic acid to a set of metabolites that contain 5 double bounds (i.e. 5-HEPE, 5-oxo-EPE, LTB5, LTC5, LTD5, and LTE5) as opposed to the 4 double bond-containing arachidonic acid metabolites. The enzyme, when acting in series with other lipoxygenase, cyclooxygenase, or cytochrome P450 enzymes, contributes to the metabolism of eicosapentaenoic acid to E series resolvins and of docosahexaenoic acid to D series resolvins. these resolvins are also classified as Specialized pro-resolving mediators.

- Arachidonate 12-lipoxygenase (ALOX12), also termed 12-lipoxygenase, platelet type platelet lipoxygenase (or 12-lipoxygenase, platelet type) 12-LOX, and 12-LO. It metabolizes arachidonic acid to 12-hydroperoxyeiocsatetraeoic acid (12-HpETE) which is further metabolized to 12-hydroxyeicosatetraenoic acid (12-HETE) or to various Hepoxilins.
- Arachidonate 15-lipoxygenase-1 (ALOX15), also termed 15-lipoxygenase-1, erythrocyte type 15-lipoxygenase (or 15-lipoxygenase, erythrocyte type), reticulocyte type 15-lipoxygenase (or 15-lipoxygenase, reticulocyte type), 15-LO-1, and 15-LOX-1. It metabolizes arachidonic acid principally to 1) 15-hydroperoxyeiocatetraenoic acid (15-HpETE) which is further metabolized to 15-Hydroxyicosatetraenoic acid (15-HETE) but also to far smaller amounts of 2) 12-hydroperoxyeicosatetraenoic acid (12-HpETE) which is further metabolized to 12-hydroxyeicosatetraenoic acid and possibly the hepoxilins. ALOX15 actually prefers linoleic acid over arachidonic acid, metabolizing linoleic acid to 12-hydroperoxyoctadecaenoic acid (13-HpODE) which is further metabolized to 13-Hydroxyoctadecadienoic acid (13-HODE). ALOX15 can metabolize polyunsaturated fatty acids that are esterified to phospholipids and/or to the cholesterol, i.e. cholesterol esters, in lipoproteins. This property along with its dual specificity in metabolizing arachidonic acid to 12-HpETE and 15-HpETE are similar to those of mouse Alox15 and has led to both enzymes being termed 12/15-lipoxygenases.
- Arachidonate 15-lipoxygenase type II (ALOX15B), also termed 15-lipoxygenase-2, 15-LOX-2, and 15-LOX-2. It metabolizes arachidonic acid to 15-hydroperoxyeicosatetraenoic (15-HpETE) which is further metabolized to 15-Hydroxyicosatetraenoic acid. ALOX15B has little or no ability to metabolize arachidonic acid to 12-hydroperoxeiocosatetraenoic acid (12-(HpETE) and only minimal ability to metabolize linoleic acid to 13-hydroperoxyoctadecaenoic acid (13-HpODE).
- Arachidonate 12-lipoxygenase, 12R type (ALOX12B), also termed 12*R*-lipoxygenase, 12*R*-LOX, and 12*R*-LO. It metabolizes arachidonic acid to 12*R*-hydroxyeicosatetraenoic acid but does so only with low catalytic activity; its most physiologically important substrate is thought to be a sphingosine which contains a very long chain (16-34 carbons) omega-hydroxyl fatty acid that is in amide linkage to the *sn*-2 nitrogen of sphingosine at its carboxy end and esterfied to linoleic acid at its

omega hydroxyl end. In skin epidermal cells, ALOX12B metabolizes the linoleate in this esterified omega-hydroxyacyl-sphingosine (EOS) to its 9*R*-hydroperoxy analog. Inactivating mutations of ALOX12B are associated with the human skin disease, autosomal recessive Congenital ichthyosiform erythroderma (ARCI).

- Epidermis-type lipoxygenase (ALOXE3), also termed eLOX3 and lipoxygenase, epidermis type. Unlike other lipoxygenases, ALOXE3 exhibits only a latent dioxygenase activity. Rather, its primary activity is as a hydroperoxide isomerase that metabolizes certain unsaturated hydroperoxy fatty acids to their corresponding epoxy alcohol and epoxy keto derivatives and thereby is also classified as a hepoxilin synthase. While it can metabolize 12*S*-hydroperoxyeicosatetraenoic acid (12*S*-HpETE) to the *R* stereoisomers of hepoxilins A3 and B3, ALOXE3 favors metabolizing *R* hydroperoxy unsaturated fatty acids and efficiently converts the 9(*R*)-hydroperoxy analog of EOS made by ALOX15B to its 9*R*(10*R*),13*R*-trans-epoxy-11*E*,13*R* and 9-keto-10*E*,12*Z* EOS analogs. ALOXE3 is thought to act with ALOX12B in skin epidermis to form the latter two EOS analogs; inactivation mutations of ALOX3 are, similar to inactivating mutations in ALOX12B, associated with autosomal recessive Congenital ichthyosiform erythroderma in humans. Inactivating mutations in ALOX3 are also associated with the human disease Lamellar ichthyosis, type 5.

Two lipoxygenases may act in series to make di-hydroxy or tri-hydroxy products that have activities quite different than either lipoxyenases' products. This serial metabolism may occur in different cell types that express only one of the two lipoxygenases in a process termed transcellular metabolism. For example, ALOX5 and ALOX15 or, alternatively, ALOX5 and ALOX12 can act serially to metabolize arachidonic acid into lipoxins while ALOX15 and possibly ALOX15B can act with ALOX5 to metabolize eicosapentaenoic acid to resolvin D's.

Mouse Lipoxygenases

The mouse is a common model to examine lipoxygenase function. However, there are some key differences between the lipoxygenases between mice and men that make extrapolations from mice studies to humans difficult. In contrast to the 6 functional lipoxygenases in humans, mice have 7 functional lipoxygenases and some of the latter have different metabolic activities than their human orthologs. In particular, mouse Alox15, unlike human ALOX15, metabolizes arachidonic acid mainly to 12-HpETE and mouse Alox15b, in contrast to human ALOX15b, is primarily an 8-lipoxygenase, metabolizing arachdionic acid to 8-HpETE; there is no comparable 8-HpETE-forming lipoxygenase in humans.

- Alox5 appears to be similar in function to human ALOX5.
- Alox12 differs from human ALOX12, which preferentially metabolizes arachidonic acid to 12-HpETE but also to substantial amounts of 15-HpETE, in that metabolizes arachidonic acid almost exclusively to 12-HpETE.

- Alox15 (also termed leukocyte-type 12-Lox, 12-Lox-l, and 12/15-Lox) differs from human ALOX15, which under standard assay conditions metabolizes arachidonic acid to 15-HpETE and 12-HpETE products in an 89 to 11 ratio, metabolizes arachidconic acid to 15-Hpete and 12-HpETE in a 1 to 6 ratio, i.e. its principal metabolite is 12-HpETE. Also, human ALOX15 prefers linoleic acid over arachidonic acid as a substrate, metabolizing it to 13-HpODE while Alox15 has little or no activity on linoleic acid. Alox15 can metabolize polyunsaturated fatty acids that are esterified to phospholipids and cholesterol (i.e. cholesterol esters). This property along with its dual specificity in metabolizing arachidonic acid to 12-HpETE and 15-HpETE are similar to those of human ALOX15 and has led to both enzymes being termed 12/15-lipoxygenases.

- Alox15b (also termed 8-lipoxygenase, 8-lox, and 15-lipoxygenase type II), in contrast to ALOX15B which metabolizes arachidonic acid principally to 15-HpETE and to a lesser extent linoleic acid to 13-HpODE, metabolizes arachidonic acid principally to 8*S*-HpETE and linoleic acid to 9-HpODE. Alox15b is as effective as ALOX5 in metabolizing 5-HpETE to leukotrienes.

- Alox12e (12-Lox-e, epidermal-type 12-Lox) is an ortholog to the human ALOX12P gene which has suffered damaging mutations and is not expressed. Alox12e prefers methyl esters over non-esterfied polyunsaturated fatty acid substrates, metabolizing linoleic acid ester to its 13-hydroperoxy counterpart and to a lesser extent arachidonic acid ester to its 12-hydroperoxy counterpart.

- Alox12b (e-LOX2, epidermis-type Lox-12) appears to act similarly to ALOX12B to metabolize the linoleic acid moiety of EOS to its 9*R*-hydroperoxy counterpart and thereby contribute to skin integrity and water impermeability; mice depleted to Alox12b develop a severe skin defect similar to Congenital ichthyosiform erythroderma. Unlike human ALOX12B which cam metabolize arachidonic acid to 12*R*-HETE at a low rate, Alox12b does not metabolize arachidonic acid as free acd but dose metabolize arachidonic acid methyl ester to its 12*R*-hydroperoxy counterpart.

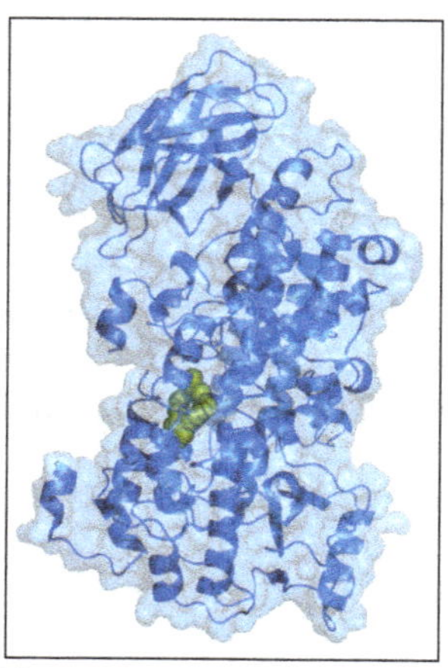

Rabbit 15-lipoxygenase (blue) with inhibitor (yellow) bound in the active site.

- Aloxe3 (epidermis-type Lox-3, eLox3) appears to act similarly to ALOXe3 in metabolizing the 9*R*-hydoperoxy-linoleate derivative of EOS to its epoxy and keto derivatives and to be involved in maintaining skin integrity and water impermeability. AloxE3 deletion leads to a defect similar to congenital ichthyosiform erythroderma.

3D Structure

There are several lipoxygenase structures known including: soybean lipoxygenase L1 and L3, coral 8-lipoxygenase, human 5-lipoxygenase, rabbit 15-lipoxygenase and porcine leukocyte 12-lipoxygenase catalytic domain. The protein consists of a small N-terminal PLAT domain and a major C-terminal catalytic domain, which contains the active site. In both plant and mammalian enzymes, the N-terminal domain contains an eight-stranded antiparallel β-barrel, but in the soybean lipoxygenases this domain is significantly larger than in the rabbit enzyme. The plant lipoxygenases can be enzymatically cleaved into two fragments which stay tightly associated while the enzyme remains active; separation of the two domains leads to loss of catalytic activity. The C-terminal (catalytic) domain consists of 18-22 helices and one (in rabbit enzyme) or two (in soybean enzymes) antiparallel β-sheets at the opposite end from the N-terminal β-barrel.

Active Site

The iron atom in lipoxygenases is bound by four ligands, three of which are histidine residues. Six histidines are conserved in all lipoxygenase sequences, five of them are found clustered in a stretch of 40 amino acids. This region contains two of the three zinc-ligands; the other histidines have been shown to be important for the activity of lipoxygenases.

The two long central helices cross at the active site; both helices include internal stretches of π-helix that provide three histidine (His) ligands to the active site iron. Two cavities in the major domain of soybean lipoxygenase-1 (cavities I and II) extend from the surface to the active site. The funnel-shaped cavity I may function as a dioxygen channel; the long narrow cavity II is presumably a substrate pocket. The more compact mammalian enzyme contains only one boot-shaped cavity (cavity II). In soybean lipoxygenase-3 there is a third cavity which runs from the iron site to the interface of the β-barrel and catalytic domains. Cavity III, the iron site and cavity II form a continuous passage throughout the protein molecule.

The active site iron is coordinated by N^{ε} of three conserved His residues and one oxygen of the C-terminal carboxyl group. In addition, in soybean enzymes the side chain oxygen of asparagine is weakly associated with the iron. In rabbit lipoxygenase, this Asn residue is replaced with His which coordinates the iron via N^{δ} atom. Thus, the coordination number of iron is either five or six, with a hydroxyl or water ligand to a hexacoordinate iron.

Details about the active site feature of lipoxygenase were revealed in the structure of porcine leukocyte 12-lipoxygenase catalytic domain complex In the 3D structure, the substrate analog inhibitor occupied a U-shaped channel open adjacent to the iron site. This channel could accommodate arachidonic acid without much computation, defining the substrate binding details for the lipoxygenase reaction. In addition, a plausible access channel, which intercepts the substrate binding channel and extended to the protein surface could be counted for the oxygen path.

LIGNIN-MODIFYING ENZYME

Lignin-modifying enzymes (LMEs) are various types of enzymes produced by fungi and bacteria that catalyze the breakdown of lignin, a biopolymer commonly found in the cell walls of plants. The terms ligninases and lignases are older names for the same class, but the name "lignin-modifying enzymes" is now preferred, given that these enzymes are not hydrolytic but rather oxidative (electron withdrawing) by their enzymatic mechanisms. LMEs include peroxidases, such as lignin peroxidase, manganese peroxidase, versatile peroxidase, and many phenoloxidases of the laccase type.

LMEs have been known to be produced by many species of white rot basidiomycetous fungi, including: Phanerochaete chrysosporium, Ceriporiopsis subvermispora, Trametes versicolor, Phlebia radiata, Pleurotus ostreatus and Pleurotus eryngii.

LMEs are produced not only by wood-white rotting fungi but also by litter-decomposing basidiomycetous fungi such as Agaricus bisporus (common button mushroom), and many Coprinus and Agrocybe species. The brown-rot fungi, which are able to colonize wood by degrading cellulose, are only able to partially degrade lignin.

Some bacteria also produce LMEs, although fungal LMEs are more efficient in lignin degradation. Fungi are thought to be the most substantial contributors to lignin degradation in natural systems.

LMEs and cellulases are crucial to ecologic cycles (for example, growth/death/decay/regrowth, the carbon cycle, and soil health) because they allow plant tissue to be decomposed quickly, releasing the matter therein for reuse by new generations of life. LMEs are also crucial to a number of different industries.

Bacterial Lignin-modifying Enzymes

Although much research has been done to understand fungal LMEs, only recently has more focus been placed on characterizing these enzymes in bacteria. The main LMEs in both fungi and bacteria are peroxidases and laccases.

Although bacteria lack homologs to the most common fungal peroxidases (lignin peroxidase, manganese peroxidase, and versatile peroxidase), many produce dye decolourizing peroxidases (DyP-type peroxidases). Bacteria from a variety of classes express DyP peroxidases, including Gammaproteobacteria, Firmicutes, and Actinobacteria. Peroxidases depolymerize lignin by oxidation using hydrogen peroxide. Fungal peroxidases have higher oxidizing power than bacterial DyP-type peroxidases studied so far, and are able to degrade more complex lignin structures. DyP-type peroxidases have been found to work on a large range of substrates, including synthetic dyes, monophenolic compounds, lignin-derived compounds, and alcohols.

Laccases, which are multicopper oxidases, are another class of enzymes found in both bacteria and fungi which have significant lignin-degrading properties. Laccases degrade lignin by oxidation using oxygen. Laccases are also widely distributed among bacterial species, including *Bacillus subtilis*, *Caulobacter crescentus*, *Escherichia coli* and *Mycobacterium tuberculosum*. Like DyP-type peroxidases, bacterial laccases have a wide substrate range.

There is interest in using bacterial laccases and DyP peroxidases for industry applications, biotechnology and bioremediation because of the greater ease of manipulation of bacterial genomes and gene expression compared to fungi. The wide range of substrates for these types of enzymes also increases the range of processes they may be used in. These processes include pulp processing, textile dye modification, decontamination of waste water and production of pharmaceutical building blocks. Furthermore, bacterial laccases function at higher temperatures, alkalinity, and salt concentrations than fungal laccases, making them more suitable for industrial use.

Both intracellular and extracellular bacterial DyP-type peroxidases and laccases have been identified, suggesting that some are used as intracellular enzymes while others are secreted to degrade compounds in the environment. However, their roles in bacterial physiology and their natural physiological substrates have yet to be detailed.

L-ISOASPARTYL METHYLTRANSFERASE

Protein L-isoaspartyl methyltransferase (PIMT, PCMT), also called S-adenosyl-L-methionine:protein-L-isoaspartate O-methyltransferase, is an enzyme which recognizes and catalyzes the repair of damaged L-isoaspartyl and D-aspartyl groups in proteins. It is a highly conserved enzyme which is present in nearly all eukaryotes, archaebacteria, and Gram-negative eubacteria.

Function

PIMT acts to transfer methyl groups from S-adenosyl-L-methionine to the alpha side chain carboxyl groups of damaged L-isoaspartyl and D-aspartyl amino acids. The enzyme takes the end methyl residue from the methionine side chain and adds it to the side chain carboxyl group of L-isoaspartate or D-aspartate to create a methyl ester. Subsequent nonenzymatic reactions result in a rapid transformation to L-succinimide, which is a precursor to aspartate and isoaspartate. The L-succinimide can then undergo nonenzymatic hydrolysis, which generates some repaired L-aspartyl residues as well as some L-isoaspartyl residues, which can then enter the cycle again for eventual conversion to the normal peptide linkage.

PIMT tends to act on proteins that have been non-enzymatically damaged due to age. By performing this repair mechanism, the enzyme helps to maintain overall protein integrity. This mechanism has been observed by several groups, and has been confirmed through experimental testing. In one report, PIMT was inhibited by adenosine dialdehyde. The results supported the proposed function of the enzyme, as the amount of abnormal L-aspartate residues increased when cells were treated with the indirect inhibitor, adenosine dialdehyde. Additionally, S-adenosylhomocysteine is known to be a competitive inhibitor of PIMT. When PIMT is not present in cells, the abnormal aspartyl residues accumulate, creating abnormal proteins that have been known to cause fatal progressive epilepsy in mice. It has been suggested that calmodulin may play a role in stimulating the function of PIMT, although the relationship between these two molecules has not been thoroughly explored. In addition to calmodulin, guanosine 5′-O-[gamma-thio]triphosphate (GTP-gammaS) has been found to stimulate PIMT activity.

Structure

The enzyme is present in human cytosol in two forms due to alternative splicing and differs among individuals in the population due to a single polymorphism at protein 119, either valine or isoleucine. The enzyme structure is described as a "doubly wound alpha/beta/alpha sandwich structure" which is quite consistent in all species analyzed thus far. If there is any difference in the sequences between different organisms it occurs in the regions connecting the three motifs in the sandwich structure, but the sequence of the individual motifs tends to be highly conserved. Researchers have found the active site to be in the loop between the beta structure and the second alpha helix and have determined it to be highly specific for isoaspartyl residues. For example, the residues found at the C-terminus of drosophila PIMT (dPIMT) are rotated 90 degrees so as to allow more space for a substrate to interact with the enzyme. In fact, dPIMT appears to alternate between this unique open conformation and the less open conformation common of PIMT in other organisms. Although possibly unrelated to this, increased levels of dPIMT in drosophila have been correlated with increase life expectancy in these organisms due to their importance in protein repair.

IMMOBILIZED ENZYME

An immobilized enzyme is an enzyme attached to an inert, insoluble material—such as calcium alginate (produced by reacting a mixture of sodium alginate solution and enzyme solution with calcium chloride). This can provide increased resistance to changes in conditions such as pH or temperature. It also lets enzymes be held in place throughout the reaction, following which they are easily separated from the products and may be used again - a far more efficient process and so is widely used in industry for enzyme catalysed reactions. An alternative to enzyme immobilization is whole cell immobilization.

Enzymes immobilised in beads of alginate gel.

Immobilization of an Enzyme

There are various ways by which one can immobilize an enzyme:

- Affinity-tag binding: Enzymes may be immobilized to a surface, e.g. in a porous material, using non-covalent or covalent Protein tags. This technology has been established for protein purification purposes. This technique is the generally applicable, and can be performed without prior enzyme purification with a pure preparation as the result. Porous glass and derivatives thereof are used, where the porous surface can be adapted in terms of hydrophobicity to suit the enzyme in question.

- Adsorption on glass, alginate beads or matrix: Enzyme is attached to the outside of an inert material. As adsorption is not a chemical reaction, the active site of the immobilized enzyme may be blocked by the matrix or bead, greatly reducing the activity of the enzyme.

- Entrapment: The enzyme is trapped in insoluble beads or microspheres, such as calcium alginate beads. However, these insoluble substances hinder the arrival of the substrate, and the exit of products.

- Cross-linkage: Enzyme molecules are covalently bonded to each other to create a matrix consisting of almost only enzyme. The reaction ensures that the binding site does not cover the enzyme's active site, the activity of the enzyme is only affected by immobility. However, the inflexibility of the covalent bonds precludes the self-healing properties exhibited by chemoadsorbed self-assembled monolayers. Use of a spacer molecule like poly(ethylene glycol) helps reduce the steric hindrance by the substrate in this case.

- Covalent bond: The enzyme is bound covalentely to an insoluble support (such as silica gel or macroporous polymer beads with epoxide groups). This approach provides the strongest enzyme/support interaction, and so the lowest protein leakage during catalysis.

Immobilization of a Substrate for Enzymatic Reactions

Another widely used application of the immobilization approach together with enzymes has been the enzymatic reactions on immobilized substrates. This approach facilitates the analysis of enzyme activities and mimics the performance of enzymes on e.g. cell walls.

INTEGRASE

Retroviral integrase (IN) is an enzyme produced by a retrovirus (such as HIV) that enables its genetic material to be integrated into the DNA of the infected cell. Retroviral INs are not to be confused with phage integrases, such as λ phage integrase (Int).

IN is a key component in the retroviral pre-integration complex (PIC). The complex of integrase bound to cognate viral DNA (vDNA) ends has been referred to as the intasome.

Structure

All retroviral IN proteins contain three canonical domains, connected by flexible linkers:

- An N-terminal HH-CC zinc-binding domain (a three-helical bundle stabilised by coordination of a Zn(II) cation).
- A catalytic core domain (RNaseH fold).
- A C-terminal DNA-binding domain (SH3 fold).

Crystal and NMR structures of the individual domains and 2-domain constructs of integrases from HIV-1, HIV-2, SIV, and Rous Sarcoma Virus (RSV) have been reported, with the first structures determined in 1994.

Biochemical data and structural data suggest that retroviral IN functions as a tetramer (dimer-of-dimers). All three domains are important for multimerisation and viral DNA binding. Early in 2010, scientists solved the crystal structure of IN from prototype foamy virus (PFV) assembled on viral DNA ends.

In addition, several host cellular proteins have been shown to interact with IN to facilitate the integration process. Human chromatin-associated protein LEDGF, which tightly binds HIV IN and directs HIV PIC towards highly expressed genes for integration, is an example of such a host factor.

Function

Integration occurs following production of the double-stranded viral DNA by the viral RNA/DNA-dependent DNA polymerase reverse transcriptase.

The main function of IN is to insert the viral DNA into the host chromosomal DNA, a step that is essential for HIV replication. Integration is a point of no return for the cell, which becomes a permanent carrier of the viral genome (provirus). Integration is in part responsible for the persistence of retroviral infections. After integration, the viral gene expression and particle production may take place immediately or at some point in the future. The timing, it is presumed, depends on the activity of the chromosomal locus hosting the provirus.

Retroviral IN catalyzes two reactions:

- 3′-processing, in which two or three nucleotides are removed from one or both 3′ ends of the viral DNA to expose the invariant CA dinucleotides at both 3′-ends of the viral DNA.
- The strand transfer reaction, in which the processed 3′ ends of the viral DNA are covalently ligated to the host chromosomal DNA.

Both reactions are catalysed by the same active site and occur via transesterification, without a covalent protein-DNA intermediate, in contrast to reactions catalysed by Ser and Tyr recombinases.

In HIV

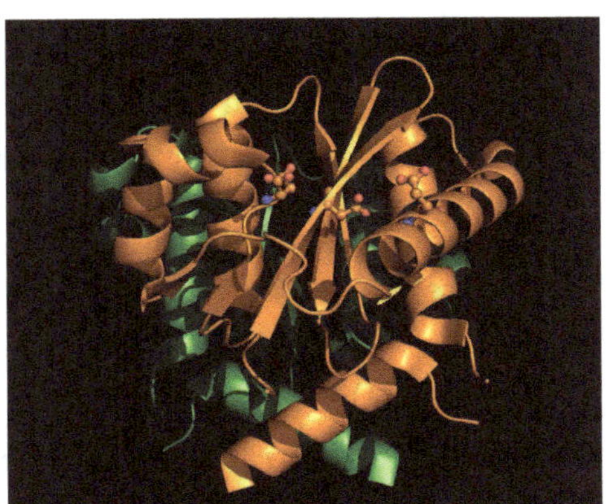

HIV Integrase shown in its full structure with its catalytic amino acids shown in ball and stick form.

HIV integrase is a 32 kDa protein produced from the C-terminal portion of the Pol gene product, and is an attractive target for new anti-HIV drugs.

In November 2005, data from a phase 2 study of an investigational HIV integrase inhibitor, MK-0518, demonstrated that the compound has potent antiviral activity. On October 12, 2007, the Food and Drug Administration (U.S.) approved the integrase inhibitor Raltegravir (MK-0518, brand name Isentress). The second integrase inhibitor, elvitegravir, was approved in the U.S. in August 2012.

The full crystal structure of human foamy virus integrase has been examined successfully. Although foamy virus is harmless to humans, the integrase is fairly similar to HIV IN and serves as a good model of it. This protein may use the morpheein model of allosteric regulation.

GLYCOGEN BRANCHING ENZYME

1,4-alpha-glucan-branching enzyme, also known as brancher enzyme or glycogen-branching enzyme is an enzyme that in humans is encoded by the *GBE1* gene.

Glycogen branching enzyme is an enzyme that adds branches to the growing glycogen molecule during the synthesis of glycogen, a storage form of glucose. More specifically, during glycogen synthesis, a glucose 1-phosphate molecule reacts with uridine triphosphate (UTP) to become UDP-glucose, an activated form of glucose. The activated glucosyl unit of UDP-glucose is then transferred to the hydroxyl group at the C-4 of a terminal residue of glycogen to form an α-1,4-glycosidic linkage, a reaction catalyzed by glycogen synthase. Importantly, glycogen synthase can only catalyze the synthesis of α-1,4-glycosidic linkages. Since glycogen is a readily mobilized storage form of glucose, the extended glycogen polymer is branched by glycogen branching enzyme to provide glycogen breakdown enzymes, such as glycogen phosphorylase, with a large number of terminal residues for rapid degradation. Branching also importantly increases the solubility and decreases the osmotic strength of glycogen.

The protein encoded by this gene is a glycogen branching enzyme that catalyzes the transfer of alpha-1,4-linked glucosyl units from the outer end of a glycogen chain to an alpha-1,6 position on the same or a neighboring glycogen chain. Branching of the chains is essential to increase the solubility of the glycogen molecule and, consequently, in reducing the osmotic pressure within cells. The highest levels of this enzyme are found in liver and muscle cells. Mutations in this gene are associated with glycogen storage disease type IV (also known as Andersen's disease).

This enzyme belongs to the family of transferases, to be specific, those glycosyltransferases that transfer hexoses (hexosyltransferases). The systematic name of this enzyme class is 1,4-alpha-D-glucan:1,4-alpha-D-glucan 6-alpha-D-(1,4-alpha-D-glucano)-transferase. Other names in common use include branching enzyme, amylo-(1,4→1,6)-transglycosylase, Q-enzyme, alpha-glucan-branching glycosyltransferase, amylose isomerase, enzymatic branching factor, branching glycosyltransferase, enzyme Q, glucosan transglycosylase, 1,4-alpha-glucan branching enzyme 1, plant branching enzyme, alpha-1,4-glucan:alpha-1,4-glucan-6-glycosyltransferase, and starch branching enzyme. This enzyme participates in starch and sucrose metabolism.

Gene

GBE is encoded by the *GBE1* gene. Through Southern blot analysis of DNA derived from human/rodent somatic cell hybrids, GBE1 has been identified as an autosomal gene located on the short arm of chromosome 3 at position 12.3. The human GBE gene was also isolated by a function complementation of the Saccharomyces cerevisiae GBE deficiency. From the isolated cDNA, the length of the gene was found to be approximately 3 kb. Additionally, the coding sequence was found to comprise 2,106 base pairs and encode a 702-amino acid long GBE. The molecular mass of human GBE was calculated to be 80,438 Da.

Structure

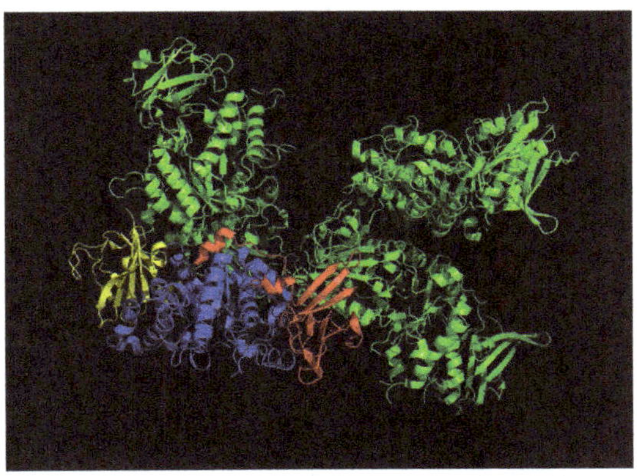

Structure of glycogen branching enzyme found in E. Coli.

Glycogen branching enzyme belongs to the α-amylase family of enzymes, which include α-amylases, pullulanas/isoamylase, cyclodextrin glucanotransferase (CGT), and branching enzyme. Shown by x-ray crystallography, glycogen branching enzyme has four marginally asymmetric units each that are organized into three domains: an amino-terminal domain, involved in determining the length of the chain transfer, a carboxyl-terminal domain, involved in substrate preference and catalytic capacity, and a central (α/β) barrel catalytic domain. The amino-terminal domain consists of 128 residues arranged in seven β-strands, the carboxyl-terminal domain with 116 residues also organized in seven β-strands, and the (α/β) barrel domain with 372 residues. While the central (α/β) barrel domain is common in members of the α-amylase family, numerous variations exist between the various barrel domains. Additionally, there are striking differences between the loops connecting elements of the secondary structure among these various α-amylase members, especially around the active site. In comparison to the other family members, glycogen binding enzyme has shorter loops, which result in a more open cavity, favorable to the binding of a bulkier substrate such as branched sugar. Through primary structure analysis and the x-ray crystallographic structures of the members of the α-amylase family, seven residue were conserved, Asp335, His340, Arg403, Asp 405, Glu458, His525, and Asp526 (E coli. numbering). These residues are implicated in catalysis and substrate binding.

Glycogen binding enzymes in other organisms have also been crystallized and structurally determined, demonstrating both similarity and variation to GBE found in Escherichia coli.

Function

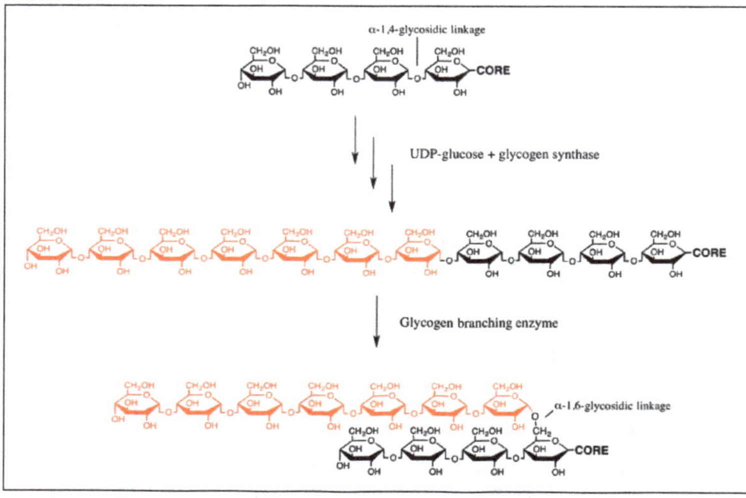

Scheme demonstrating the function of glycogen branching enzyme.

In glycogen, every 10 to 14 glucose units, a side branch with an additional chain of glucose units occurs. The side chain attaches at carbon atom 6 of a glucose unit, an α-1,6-glycosidic bond. This connection is catalyzed by a branching enzyme, generally given the name α-glucan branching enzyme. A branching enzyme attaches a string of seven glucose units (with some minor variation to this number) to the carbon at the C-6 position on the glucose unit, forming the α-1,6-glycosidic bond. The specific nature of this enzyme means that this chain of 7 carbons is usually attached to a glucose molecule that is in position three from the non-reducing end of another chain. Because the enzyme works with such specificity regarding the number of glucose units transferred and the position to which they are transferred, the enzyme creates the very characteristic, highly branched glycogen molecule.

Clinical Significance

Mutations in this gene are associated with glycogen storage disease type IV (also known as Andersen's disease) in newborns and with adult polyglucosan body disease.

Approximately 40 mutations in the GBE1 gene, most resulting in a point mutation in the glycogen branching enzyme, have led to the early childhood disorder, glycogen storage disease type IV (GSD IV). This disease is characterized by a severe depletion or complete absence of GBE, resulting in the accumulation of abnormally structured glycogen, known as polyglucosan bodies. Glycogen buildup leads to increased osmotic pressure resulting in cellular swelling and death. The tissues most affected by this disease are the liver, heart, and neuromuscular system, areas with the greatest levels of glycogen accumulation. Abnormal glycogen buildup in the liver interferes with liver functioning and can result in an enlarged liver and liver disease. In muscles, the inability of cells to efficiently breakdown glycogen due to the severe reduction or absence of branching can lead to muscle weakness and atrophy. At least three mutations in the GBE1 gene have been found to cause another disease called adult polyglucosan body disease (APBD). While in GSD IV GBE activity is undetectable or minimally detectable, APBD is characterized by reduced or even normal GBE activity. In this disease, abnormal glycogen can build up in neurons leading to a spectrum of problems. Specifically, some disease characteristics are gait difficulties from mixed upper and lower motor neuron involvement sensory loss in lower extremities, and neurogenic bladder, a problem in which a person lacks bladder control due to a brain, spinal cord, or nerve condition.

GLUCANSUCRASE

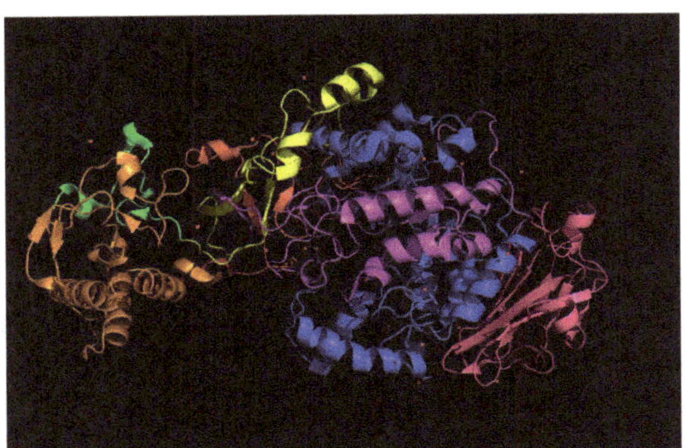

Glucansucrase in Streptococcus mutans. The domains are color coded. For domains made up of discontiguous segments, each segment was assigned a number. Shown here are segments IV1 (orange), B1 (red), A1 (blue), C (pink), A2 (purple), B2 (yellow), and IV2 (green).

Glucansucrase (also known as glucosyltransferase) is an enzyme in the glycoside hydrolase family GH70 used by lactic acid bacteria to split sucrose and use resulting glucose molecules to build long, sticky biofilm chains. These extracellular homopolysaccharides are called α-glucan polymers.

Glucansucrase enzymes can synthesize a variety of glucans with differing solubilities, rheology, and other properties by altering the type of glycosidic linkage, degree of branching, length, mass, and conformation of the polymers. Glucansucrases are classified according to the glycosidic linkage they catalyze. They can be mutansucrases, dextransucrases, alternansucrases, or reuteransucrases. This versatility has made glucansucrase useful for industrial applications. Glucansucrase's role in cariogenesis is a major point of interest. Glucan polymers stick to teeth in the human mouth and cause tooth decay.

Structure

Glucansucrases are large, extracellular proteins with average molecular masses around 160,000 Daltons. Therefore crystallography studies have only been carried out for fragments of the enzymes, not complete structures. However, glucansucrase is very similar to α-amylase, another sugar-cutting enzyme. Glucansucrase thus has many of the same structural features. For example, both enzymes have three domains in their catalytic core and a $(\beta/\alpha)_8$ barrel.

Glucansucrase has 5 major domains: A, B, C, IV, and V. The domains in glucansucrase, however, have a different arrangement than those in α-amylase. The folding characteristics of α-amylase and glucansucrase are still very similar, but their domains are permuted. Domains A, B, IV, and V are built from two discontiguous parts of the polypeptide chain, causing the chain to follow a U-shape. From the N- to C-terminus, the polypeptide chain goes in the following order: V, IV, B, A, C, A, B, IV, V. The C domain is the only one made up of a continuous polypeptide sequence.

Domain A contains the $(\beta/\alpha)_8$ barrel and the catalytic site. In the catalytic site, three residues in particular play important roles for enzymatic activity: a nucleophilic aspartate, an acid/base glutamate, and an additional aspartate to stabilize the transition state.

Domain B makes up a twisted antiparallel β sheet. Some of the loops in domain B help shape the groove near the catalytic site. Additionally, some amino acids between domains A and B form a calcium binding site near the nucleophilic aspartate. The Ca^{2+} ion is necessary for enzyme activity.

Reaction and Mechanism

Glucansucrase has two parts to its reaction. First it cleaves a glycosidic bond to split sucrose. Products of the reaction are the constituent monosaccharides glucose and fructose. This glucose is added to a growing glucan chain. Glucansucrase uses the energy released from bond cleavage to drive glucan synthesis. Both sucrose breakdown and glucan synthesis occur in the same active site.

The first step is carried out through a transglycosylation mechanism involving a glycosyl-enzyme intermediate in subsite-1. Glutamate is likely the catalytic acid/base, aspartate the nucleophile, and another aspartate the transition state stabilizer. These three residues are all highly conserved and mutating them leads to a significant decrease in enzymatic activity.

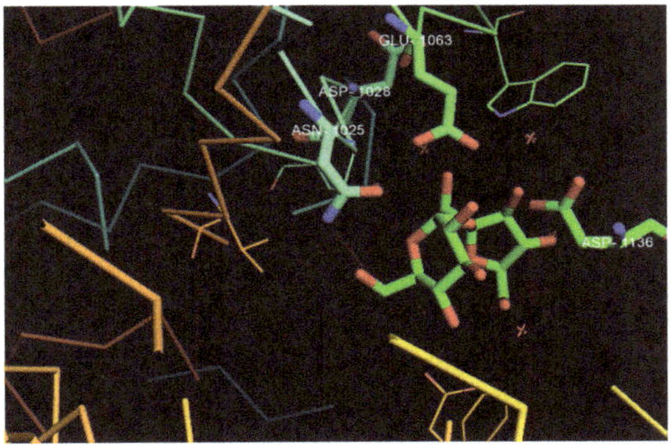

Active site of glucansucrase in Lactobacillus reuteri

The glucansucrase mechanism has historically been controversial in the scientific literature. The mechanism involves two displacements. The first originates from a glycosidic cleavage of the sucrose substrate between subsites -1 and +1. This releases fructose and forms a sugar-enzyme intermediate when the glucose unit attaches to the nucleophile.

The second displacement is transfer of a glucosyl moiety to an acceptor, such as a growing glucan chain. The debate in the past was over whether the glucosyl group attached to the non-reducing or reducing end of an incoming acceptor. Additional investigations pointed to a non-reducing mechanism with a single active site.

Evolution

Glucansucrase proteins likely evolved from an amylase enzyme precursor. The two enzymes have similar folding patterns and protein domains. In fact, past attempts to produce drugs targeting glucansucrase have not been successful because the drugs also disrupted amylase, which is necessary to break down starches. This occurred because the active sites of the two enzymes are nearly the same. Glucansucrase likely maintained a highly-conserved active site as it underwent a different evolutionary path.

Health

Glucansucrase allows the oral bacteria *Streptococcus mutans* to metabolize sucrose into lactic acid. This lactic acid lowers the pH around teeth and dissolves calcium phosphate in tooth enamel, leading to tooth decay. Additionally, the synthesis of glucan aids *S. mutans* in adhering to the surface of teeth. As the polymers accumulate, they help more acid-producing bacteria stay on teeth. Consequently, glucansucrase is such an attractive drug target to prevent tooth decay. If *S. mutans* can no longer break down sucrose and synthesize glucan, calcium phosphate is not degraded and bacteria cannot adhere as easily to teeth.

EPOXYGENASE

Epoxygenases are a set of membrane-bound, heme-containing cytochrome P450 (CYP P450 or just CYP) enzymes that metabolize polyunsaturated fatty acids to epoxide products that have a range of biological activities. The most thoroughly studied substrate of the CYP epoxylgenases is arachidonic acid. This polyunsaturated fatty acid is metabolized by cyclooxygenases to various prostaglandin, thromboxane, and prostacyclin metabolites in what has been termed the first pathway of eicosanoid production; it is also metabolized by various lipoxygenases to hydroxyeicosatetraenoic acids (e.g. 5-Hydroxyeicosatetraenoic acid, 12-Hydroxyeicosatetraenoic acid, 15-hydroxyicosatetraenoic acid) and leukotrienes (e.g. leukotriene B4, leukotriene C4) in what has been termed the second pathway of eicosanoid production. The metabolism of arachidonic acid to epoxyeicosatrienoic acids by the CYP epoxygenases has been termed the third pathway of eicosanoid metabolism. Like the first two pathways of eicosanoid production, this third pathway acts as a signaling pathway wherein a set of enzymes (the epoxygenases) metabolize arachidonic acid to a set of products (the eicosatrienoic acid epoxides, abbreviated as EETs, which are classified as nonclassic eicosanoids) that act as secondary signals to work in activating their parent or nearby cells and thereby orchestrate functional responses.

However, none of these three pathways is limited to metabolizing arachidonic acid to eicosanoids. Rather, they also metabolize other polyunsaturated fatty acids to products that are structurally analogous to the eicosanoids but often have different bioactivity profiles. This is particularly true for the CYP epoxygenases which in general act on a broader range of polyunsaturated fatty acids to form a broader range of metabolites than the first and second pathways of eicosanoid production. Furthermore, the latter pathways form metabolites many of which act on cells by binding with and thereby activating specific and well-characterized receptor proteins; no such receptors have been fully characterized for the epoxide metabolites. Finally, there are relatively few metabolite-forming lipoxygenases and cyclooxygenases in the first and second pathways and these oxygenase enzymes share similarity between humans and other mammalian animal models. The third pathway consists of a large number of metabolite-forming CYP epoxygenases and the human epoxygenases have important differences from those of animal models. Partly because of these differences, it has been difficult to define clear roles for the epoxygenase-epoxide pathways in human physiology and pathology.

CYP Epoxygenases

The cytochrome P450 (CYP) superfamily of membrane-bound (typically endoplasmic reticulum-bound) enzymes contain a heme cofactor and therefore are hemoproteins. The superfamily comprises more than 11,000 genes categorized into 1,000 families that are distributed broadly throughout bacteria, archaea, fungi, plants, animals, and even viruses. The CYP enzymes metabolize an enormously large variety of small and large molecules including foreign chemical substances, i.e. xenobiotics and pharmaceuticals, as well as a diversity of endogenously formed substances such as various steroids, vitamin D, bilirubin, cholesterol, and fatty acids. Humans have 57 putatively active CYP genes and 58 CYP pseudogenes of which only a few are polyunsaturated fatty acid (PUFA) epoxygenases, i.e. enzymes with the capacity to attach atomic oxygen to the carbon-carbon double bonds of long chain PUFA to form their corresponding epoxides. These CYP epoxygenases represent a family of enzymes that consists of several members of the CYP1 and CYP2 subfamilies. The metabolism of the straight chain 20-carbon polyunsaturated fatty eicosatetraenoic acid, arachidonic acid, by certain CYP epoxygenases is a good example of their action. Arachidonic acid has 4 cis-configured double bonds located between carbons 5-6, 8-9, 11-12, and 14-15 Double bonds. It is therefore 5*Z*,8*Z*,11*Z*,14*Z*-eicosatetraenoic acid. Certain CYP epoxygenases attack these double bounds to form their respective eicosatrienoic acid epoxide regioisomers. The products are therefore 5,6-EET (i.e. 5,6-epoxy-8*Z*,11*Z*,14*Z*-eicosatetrienoic acid), 8,9-EET (i.e. 5,6-epoxy-8*Z*,11*Z*,14*Z*-eicosatetrienoic acid), 11,12-EET (i.e. 11,12-epoxy-5*Z*,8*Z*,14*Z*-eicosatetrienoic acid), and/or 14,15-EET.

Note that the eicosatetraenoate substrate loses one double bound to become an eicosatrienoic acid with three double bonds and that the epoxygenases typically form a mixture of *R*/*S* enantiomers at the attacked double bound position. Thus, the CYP epoxygenases which attack arachidonic acid's double bound between carbon 14 and 15 form a mixture of 14*R*,15*S*-ETE and 14*S*,15*R*-ETE. However, each CYP epoxygenase often shows preferences in the position of the double bound on which they act, partial selectivity in the *R*/*S* enantiomer ratios that they make at each double bound position, and different double bond position preferences and *R*/*S* selectivity ratios with different PUFA substrates. Finally, the product epoxides are short-lived in cells, generally existing for only several seconds before being converted by a Soluble epoxide hydrolase (also termed epoxide hydrolase 2 or sEH) to their corresponding dihydroxy-eicosatetraenoic acid (diHETE) products, e.g. 14,15-HETE rapidly becomes a mixture of 14(*S*),15(*R*)-diHETE and 14(*R*),15(*S*)-diHETE. Although there are exceptions, the diHETE products are generally far less active than their epoxide precursors; the sEH pathway is therefore regarded as an inactivating pathway which functions to limit epoxide activity.

The catalytic activity of endoplasmic reticulum-bound cytochrome P450 enzymes, including the epoxygenases, depends upon Cytochrome P450 reductase (POR); it transfers electrons to, and thereby regenerates the activity of, the CYPs. The human gene that expresses POR is highly polymorphic; many of the polymorphic variant PORs cause significantly decreases or increases in the activity of the CYPs, including the epoxygenases.

Scores of drugs have been shown to either inhibit or induce one or more of the CYP epoxygenases.

CYP Epoxygenase Substrates and Products

The most studied substrate of the CYP epoxygenases is the omega-6 fatty acid, arachidonic acid. However, the CYP epoxygenases also metabolize other omega-6 fatty acids such as linoleic acid and the omega-3 fatty acids, eicosapentaenoic acid and docosahexaenoic acid. The distinction between the omega-6 and omega-3 fatty acid substrates is important because omega-3 fatty acids metabolites can have lesser or different activities than omega-6 fatty acid metabolites; furthermore, they compete with the omega-6 fatty acids for the CYP epoxygenases thereby reducing the production of omega-6 fatty acid metabolites. The human CYP P450 enzymes identified to have epoxygenase activity on one or more PUFA include CYP1A1, CYP1A2, CYP2C8, CYP2C9, CYP2C18, CYP2C19, CYP2E1, CYP2J2, CYP2S1, CYP3A4, CYP4F2, CYP4F3A, CYP4F3B, CYP4A11, CYP4F8, and CYP4F12. CYP2C8 and CYP2C9 form particularly large amounts of superoxide anion (chemical formula O_2^-) during their metabolism of polyunsaturated fatty acids; this reactive oxygen species is toxic to cells and may be responsible for some of the activities ascribed to the epoxides made by the two CYPs.

Omega-6 Fatty Acids

Arachidonic Acid

In humans, CYP1A1, CYP1A2, CYP2C8, CYP2C9, CYP2C18, CYP2C19, CYP2E1, CYP2J2, and CYP2S1 isoforms metabolize arachidonic acid to Epoxyeicosatrienoic acids (i.e., EETs) as defined using recombinient CYPs in an *In vitro* microsome assay. CYP2C9, and CYP2J2 appear to be the main producers of the EETs in humans with CYPP2C9 being the main unsaturated fatty acid epoxide producer in vascular endothelial cells and CYP2J2 being highly expressed (although less catalytically active than CYP2C9) particularly in heart muscle but also in kidneys, pancreas, lung, and brain; CYP2C8, CYP2C19, CYP2J2 are also implicated in converting arachidonic acid to epoxides in humans. Most of these CYPs preferentially form 14,15-ETE, somewhat lower levels of 11,12-EET, and far lower, trace, or undetectable levels of 8,9-ETE and 4,5-ETE. There are exceptions to this rule with, for example, CYPE1 forming 14,15-EET almost exclusively, CYP2C19 forming 8,9-EET at slightly higher levels than 14,15-EET, and CYP3A4 forming 11,12-EET at slightly higher levels than 14,15-ETE. 14,15-EET and 11,12-EET are the major EETs produced by mammalian, including human, tissues. CYP2C9, CYP2JP, and possibly the more recently characterized CYP2S1 appear to be the main produces of the EETs in humans with CYPP2C9 being the main EET producer in vascular endothelial cells and CYP2JP being highly expressed (although less catalytically active than CYP2C) in heart muscle, kidneys, pancreas, lung, and brain. CYP2S1 is expressed in macrophages, liver, lung, intestine, and spleen and is abundant in human and mouse atherosclerosis (i.e. Atheroma) plaques as well as inflamed tonsils. CYP2S1 is expressed in macrophages, liver, lung, intestine, and spleen; is abundant in human and mouse atherosclerosis (i.e. Atheroma) plaques as well as inflamed tonsils; and, in addition to forming epoxides of arachidonic acid (and other polyunsaturated fatty acids), CYP2S1 metabolizes prostaglandin G2 and Prostaglandin H2 to 12-Hydroxyheptadecatrienoic acid. Possibly because of metabolizing and thereby inactivating the prostaglandins and/or because forming the bioactive metabolite, 12-hyddroxyheptadecatrienoic acid, rather than EETs, CYP2S1 may act to inhibit the function of monocytes and thereby limit inflammation as well as other immune responses.

Linoleic Acid

CYP2C9 and CYP2S1 are known to, and many or all of the other CYPs that act on arachidonic acid are thought to, metabolize the 18 carbon essential fatty acid, 9(*Z*),12(*Z*)-octadecadienoic acid, i.e. linoleic acid, at is 12,13 carbon-carbon double bout to form (+) and (-) epoxy optical isomers viz., the 9*S*,10*R*-epoxy-12(*Z*)-octadecaenoic and 9*R*,10*S*-epoxy-12(*Z*)-octadecaenoic acids; this set of optical isomers is also termed vernolic acid, linoleic acid 9:10-oxide, and leukotoxin. CYPC2C9 is known and the other arachidonic acid-metabolizing CYPs are thought to likewise attack linoleic acid at its 9,10 carbon-carbon double bound to form 12*S*,13*R*-epoxy-9(*Z*)-octadecaenoic and 12*R*,13*S*-epoxy-9(*Z*)-octadecaenoic acid optical isomers; this set of optical isomers is also termed coronaric acid, linoleic acid 12,13-oxide, and isoleukotoxin These linoleic acid-derived leukotoxin and isoleukotoxin sets of optical isomers possess activities similar to that of other leukotoxins such as the pore-forming leukotoxin family of RTX toxin virulence factor proteins secreted by gram-negative bacteria, e.g. Aggregatibacter actinomycetemcomitans and E. coli. That is, they are toxic to leukocytes as well as many other cell types and when injected into rodents produce multiple organ failure and respiratory distress. These effects appear due to the conversion of leukotoxin to its dihydroxy counterparts, 9*S*,10*R*- and 9*R*,10*S*-dihydroxy-12(*Z*)-octadecaenoic acids, and isoleukotoxin to its 12*R*,13*S*- and 12*S*,13*R*-dihydroxy-9(*Z*)-octadecenoic acid counterparts by soluble epoxide hydrolase. Some studies suggest but have not proven that leukotoxin and isoleukotoxin, acting primarily if not exclusively through their respective dihydroxy counterparts, are responsible for or contribute to multiple organ failure, respiratory distress, and certain other cataclysmic diseases in humans.

Adrenic Acid

Adrenic acid or 7(*Z*),10(*Z*),13(*Z*),16(*Z*)-docosatetraenoic acid, an abundant fatty acid in the adrenal gland, kidney, vasculature, and early human brain, is metabolized primarily to 7(*Z*),10(*Z*),13(*Z*)-16,17-epoxy-docosatrienoic acid and smaller amounts of its 7,8-, 10,11-, and 13,14-epoxide-docosatrienoic acids by bovine coronary arteries and adrenal zona glomerulosa cells through the apparent action of an unidentified CYP epoxygenase(s); the eSH-dependent metabolism of these eoxide, 7,8-, 10,11-, and 13,14-dihydroxy-docosatrienoic acids relaxes pre-contracted coronary and adrenal gland arteries suggesting that the dihydroxy metabolites may act as vascular endothelium-derived Endothelium-derived relaxing factors.

Omega-3 Fatty Acids

Eicosapentaenoic Acid

5(*Z*),8(*Z*),11(*Z*),14(*Z*),17(*Z*)-eicosapentaenoic acid (EPA) is metabolized by the same CYP epoxygenases that metabolize arachidonic acid primarily to 17, 18-epoxy-5(*Z*),8(*Z*),11(*Z*),14(*Z*)-eicosatetranoic acid and usually far smaller or undetectable amounts of EPA's 5,6-, 8,9-, 11,12-, or 14,15-epoxides; however, CYP2C9 metabolizes EPA primarily to 14,15-epoxy-5(*Z*),8(*Z*),11(*Z*),17(*Z*)-eicosatetranoic acid, CYP2C11 forms appreciable amounts of this 14,15-epoxide in addition to the 17,18-epoxide, and CYP2C18 forms appreciable amounts of the 11,12 epoxide (11,12-epoxy-5(*Z*),8(*Z*),14(*Z*),17(*Z*)-eicosatetranoic acid) in addition to the 17,18-epoxide. Furthermore, CYP4A11, CYP4F8, and CYP4F12, which are CYP monooxygenase rather than CYP epoxygeanse in that they metablize arachidonic acid to monohydroxy eicosatetraenoic acid products, i.e. 19-hydroxy- and/or 18-hydroxy-eicosatetranoic acids, takes on epoxygease activity in converting EPA primarily to its 17,18-epoxy metabolite.

Docosahexaenoic Acid

4(*Z*),7(*Z*),10(*Z*),13(*Z*),16(*Z*),19(*Z*)-docosahexaenoic acid (DHA) is metabolized by the same CYP eoxygenases that metabolize arachidonic acid to form epoxide-containing docosapentaenoic acid products, particularly 19,20-epoxy-4(*Z*),7(*Z*),10(*Z*),13(*Z*),16(*Z*)-docosapentenoic acid. These docosapentaenoic acid epoxides or Epoxydocosapentaenoic acids (EDPs) have a somewhat different set of activities than, and thereby may serve in part as counterpoises to, the EET's; EDPs may also be responsible for some the beneficial effects attributed to omega-6 fatty acid-rich foods such as fish oil.

α-Linolenic Acid

The 18 carbon essential fatty acid, 9 α-Linolenic acid or 9(*Z*),12(*Z*),15(*Z*)-octadecatrienoic acid, is metabolized primarily to 9(*Z*),12(*Z*)-15,16-epoxy-octadecadienoic acid but also to smaller amounts of its 8,10- and 12,13-epoxides in the serum, liver, lung, and spleen of mice treated with a drug that increases the expression of CYP1A1, CYP1A2 and/or CYP1B1. These epoxides are also found in the plasma of humans and their levels greatly increase in subjects given an α-linolenic acid-rich diet.

Genetic Polymorphism in CYP Epoxygenases

Human CYP epoxygenase genes come in many single nucleotide polymorphism (SNP) variants some of which code for epoxygenase products with altered activity. Investigation into the impact of these variants on the bearers' health (i.e. phenotype) is an invaluable area of research which offers the opportunity to define the function of the epoxygenases and their polyunsaturated fatty acid metabolites in humans. However, SNP variants that cause altered polyunsaturated fatty acid metabolism may also cause altered metabolism of their other substrates, i.e. diverse xenobiotic (e.g. NSAID) and endotiotic (e.g. the primary female sex hormone, estradiol) compounds: the latter effects may lead to clinical manifestations that overshadow any manifestations resulting from changes in polyunsaturated fatty acid metabolism.

The most common SNP epoxygenase variants are as follows:

1. CYP2C8*3 (30411A>G, rs10509681, Lys399Arg) converts arachidonic acid to 11,12-EET and 14,15-EET with a turnover rate less than half that of wild type CYP2C8; in a single recent report, male but not female carriers of the CYP2C8*3 allele had an increased risk

of essential hypertension. Bea rers of this SNP may or may not show increased risk of developing acute gastrointestinal bleeding during the use of non-steroidal anti-inflammatory drugs (NSAIDs) that are its substrates such as aceclofenac, celecoxib, diclofenac, ibuprofen, indomethacin, lornoxicam, meloxicam, naproxen, piroxicam, tenoxicam, and valdecoxib.

2. CYP2J2*7 (−76G>T, rs890293, upstream Promoter (genetics) site) has decreased binding of the Sp1 transcription factor resulting in its lowered expressionas and lowered levels of EETs in plasma. Carriers of this SNP among a Uyghur population in China had a higher risk of coronary artery disease. However, CYP2J2*7 carriers showed no association with hypertension, heart attack, or stroke in a study of 5,740 participants of the cardiovascular cohort of the Malmö Diet and Cancer study; since other studies have afforded contradictory results, this allele is currently regarded as not associated with cardiovascular diseases. Bearers of this SNP in a Chinese Population had a higher risk of younger onset of type 2 diabetes and among a Chinese Han population had a higher risk of Alzheimer's disease.

3. CYP2C8*2 (11054A>T, rs11572103, Ile269Phe) and CYP2C8*4 (11041C>, rs1058930, variants have reduced arachidonic acid-metabolizing activity but have not been clearly associated with cardiovascular or other diseases.

4. CYPC28*4 (3608C>T, rs1058930, Ile264Met) has reduced arachidonic acid metabolizing activity. It has not been associated with cardiovascular diseases but has a higher incidence in subjects with type II diabetes in a small sample of Caucasians in Germany.

5. The CYP2C9*2 (3608C>T, rs1799853, Arg144Cys) variant has a 50% reduction in polyunsaturated fatty acid metabolizing activity compared to CYP2C9 wild type; carriers of it show no association with cardiovascular disease but exhibit poor metabolism of the anti-coagulating, *blood-thining* agent, warfarin. These carriers are susceptible to the gastrointestinal bleeding side effects of warfarin and the NSAID cited above.

6. CYP2C9*3 (42624A>C, rs1057910, Iso359Leu) encodes an expoxygenase with reduced arachidonic acid metabolizing activity. This allele has not been directly associated with cardiovascular diseases but may be associated with the poor metabolism and therefore adverse reactions to warfarin, NSAID, sulfonylurea-containing oral hypoglycemic agents, and the anti-(epilepsy) drug, phenytoin.

7. CYP2C19*2 (19154G>A, rs4244285, Il264Met) and CYP2C19*3 (17948G>A, rs4986893, His212X) are loss-of-function null alleles; carriers of the CYP2C19*3 but not the CYP2C19*2 allele showed a *reduced* risk of developing essential hypertension in a large Korean population study. Bearers of null alleles would be expected to be poor metabolizers of several drugs which are CYP2C19*2 or CYP2C19*3 substrates. This is particularly the case with Clopidogrel, a drug used to block platelet activation, blood clotting, and thereby heart attack, stroke, and peripheral artery occlusion in people at high risk of these events; CYP2C19 metabolizes clopidogrel to it active form. Consequently, patients with severe deficiencies in this CYP, i.e. bearers of CYP2C19*3 or CYP2C19*2 alleles, fail to gain protection from clopidogrel and have a higher risk of the cited cardiovascular events than clopidogrel-treated patients bearing wild type CYP2C19 alleles.

8. CYPC19*17 (-800C>T, rs12248560, site upstream gene promoter site) causes overproduction of its epoxygenase and thereby the ultra fast metabolism of arachidonic acid. Bearers of this allele have not been associated with cardiovascular diseases but clearly show a decreased risk of developing breast cancer and endometriosis possible because their rapid metabolism of estrogen leads to lower estrogen levels and thereby a lower risk of these estrogen-fueled diseases. These bearers also a higher rate of metabolism of, and therefore reduced responsiveness to, certain proton pump inhibitor and antidepressant drugs.

Genetic Polymorphism in Cytochrome P450 Reductase

As indicated above, Cytochrome P450 reductase (POR) is responsible for regenerating the activity of CYPs including the epoxygenases. Several genetic variants of the human POR gene impact epoxygenase activity. For example, POR Missense mutations A287P and R457H lead to reductions in the activity of CYP2C19 and CYP2C9, respectively, whereas A503V and Q153R missense mutations lead to small increases in the activity of CYP2C9. While these and other POR genetic variants have not yet been associated with epoxygenase-related disease, they contribute to the marked variability in the activity of the epoxygenases between individuals.

CELLULASE

Cellulase is any of several enzymes produced chiefly by fungi, bacteria, and protozoans that catalyze cellulolysis, the decomposition of cellulose and of some related polysaccharides. The name is also used for any naturally occurring mixture or complex of various such enzymes, that act serially or synergistically to decompose cellulosic material.

Cellulases break down the cellulose molecule into monosaccharides ("simple sugars") such as beta-glucose, or shorter polysaccharides and oligosaccharides. Cellulose breakdown is of considerable economic importance, because it makes a major constituent of plants available for consumption and use in chemical reactions. The specific reaction involved is the hydrolysis of the 1,4-beta-D-glycosidic linkages in cellulose, hemicellulose, lichenin, and cereal beta-D-glucans. Because cellulose molecules bind strongly to each other, cellulolysis is relatively difficult compared to the breakdown of other polysaccharides such as starch.

Most mammals have only very limited ability to digest dietary fibres such as cellulose by themselves. In many herbivorous animals such as ruminants like cattle and sheep and hindgut fermenters like horses, cellulases are produced by symbiotic bacteria. Cellulases are produced by a few types of animals, such as some termites, snails, and earthworms.

Several different kinds of cellulases are known, which differ structurally and mechanistically. Synonyms, derivatives, and specific enzymes associated with the name "cellulase" include endo-1,4-beta-D-glucanase (beta-1,4-glucanase, beta-1,4-endoglucan hydrolase, endoglucanase D, 1,4-(1,3,1,4)-beta-D-glucan 4-glucanohydrolase), carboxymethyl cellulase (CMCase), avicelase, celludextrinase, cellulase A, cellulosin AP, alkali cellulase, cellulase A 3, 9.5 cellulase, and pancellase SS. Enzymes that cleave lignin have occasionally been called cellulases, but this old usage is deprecated; they are lignin-modifying enzymes.

Types and Action

Five general types of cellulases based on the type of reaction catalyzed:

- Endocellulases randomly cleave internal bonds at amorphous sites that create new chain ends.
- Exocellulases or cellobiohydrolases cleave two to four units from the ends of the exposed chains produced by endocellulase, resulting in tetrasaccharides or disaccharides, such as cellobiose. Exocellulases are further classified into type I, that work processively from the reducing end of the cellulose chain, and type II, that work processively from the nonreducing end.
- Cellobiases or beta-glucosidases hydrolyse the exocellulase product into individual monosaccharides.
- Oxidative cellulases depolymerize cellulose by radical reactions, as for instance cellobiose dehydrogenase (acceptor).
- Cellulose phosphorylases depolymerize cellulose using phosphates instead of water.

Avicelase has almost exclusively exo-cellulase activity, since avicel is a highly micro-crystalline substrate.

Within the above types there are also progressive (also known as processive) and nonprogressive types. Progressive cellulase will continue to interact with a single polysaccharide strand, nonprogressive cellulase will interact once then disengage and engage another polysaccharide strand.

Cellulase action is considered to be synergistic as all three classes of cellulase can yield much more sugar than the addition of all three separately. Aside from ruminants, most animals (including humans) do not produce cellulase in their bodies and can only partially break down cellulose through fermentation, limiting their ability to use energy in fibrous plant material.

Structure

Most fungal cellulases have a two-domain structure, with one catalytic domain and one cellulose binding domain, that are connected by a flexible linker. This structure is adapted for working on an insoluble substrate, and it allows the enzyme to diffuse two-dimensionally on a surface in a caterpillar-like fashion. However, there are also cellulases (mostly endoglucanases) that lack cellulose binding domains.

Both binding of substrates and catalysis depend on the three-dimensional structure of the enzyme which arises as a consequence of the level of protein folding. The amino acid sequence and arrangement of their residues that occur within the active site, the position where the substrate binds, may influence factors like binding affinity of ligands, stabilization of substrates within the active site and catalysis. The substrate structure is complementary to the precise active site structure of enzyme. Changes in the position of residues may result in distortion of one or more of these interactions. Additional factors like temperature, pH and metal ions influence the non-covalent interactions between enzyme structure. The Thermotoga maritima species make cellulases consisting of 2 beta-sheets (protein structures) surrounding a central catalytic region which is the active-site. The enzyme is categorised as an endoglucanase, which internally cleaves β-1,4 -glycosydic bonds in cellulose chains facilitating further degradation of the polymer. Different species in the same family as T. Maritima make cellulases with different structures. Cellulases produced by the species Coprinopsis Cinerea consists of seven protein strands in the shape of an enclosed tunnel called a beta/alpha barrel. These enzymes hydrolyse the substrate carboxymethyl cellulose. Binding of the substrate in the active site induces a change in conformation which allows degradation of the molecule.

Cellulase Complexes

In many bacteria, cellulases in-vivo are complex enzyme structures organized in supramolecular complexes, the cellulosomes. They can contain, but are not limited to, five different enzymatic subunits representing namely endocellulases, exocellulases, cellobiases, oxidative cellulases and cellulose phosphorylases wherein only endocellulases and cellobiases participate in the actual hydrolysis of the $\beta(1\rightarrow 4)$ linkage. The number of sub-units making up cellulosomes can also determine the rate of enzyme activity.

The cellulase complex from *Trichoderma reesei*, for example, comprises a component labeled C1 (57,000 daltons) that separates the chains of crystalline cellulose, an endoglucanase (about 52,000 daltons), an exoglucanase (about 61,000 dalton), and a beta-glucosidase (76,000 daltons).

Numerous "signature" sequences known as dockerins and cohesins have been identified in the genomes of bacteria that produce cellulosomes. Depending on their amino acid sequence and tertiary structures, cellulases are divided into clans and families.

Mechanism of Cellulolysis

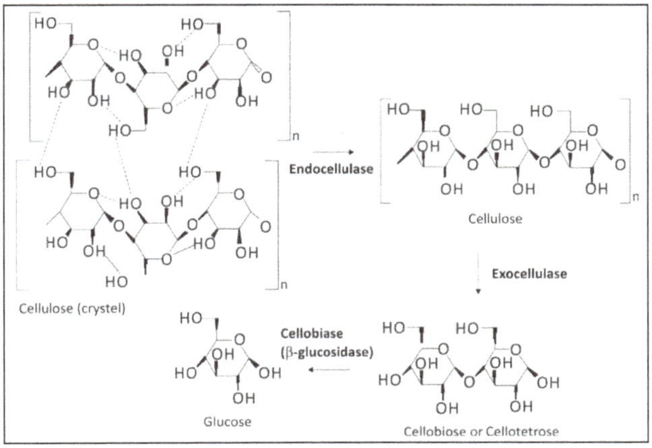

The three types of reaction catalyzed by cellulases:1. Breakage of the noncovalent interactions present in the amorphous structure of cellulose (endocellulase) 2. Hydrolysis of chain ends to break the polymer into smaller sugars (exocellulase) 3. Hydrolysis of disaccharides and tetrasaccharides into glucose (beta-glucosidase).

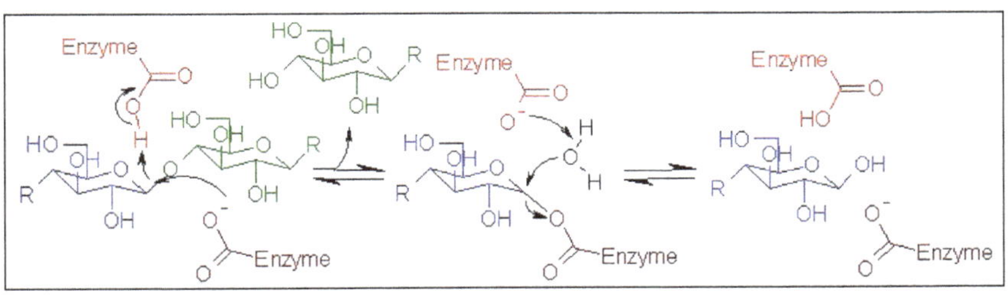

Mechanistic details of beta-glucosidase activity of cellulase.

Uses

Cellulase is used for commercial food processing in coffee. It performs hydrolysis of cellulose during drying of beans. Furthermore, cellulases are widely used in textile industry and in laundry detergents. They have also been used in the pulp and paper industry for various purposes, and they are even used for pharmaceutical applications. Cellulase is used in the fermentation of biomass into biofuels, although this process is relatively experimental at present. Medically, Cellulase is used as a treatment for phytobezoars, a form of cellulose bezoar found in the human stomach, and it has exhibited efficacy in degrading polymicrobial bacterial biofilms by hydrolyzing the β(1-4) glycosidic linkages within the structural, matrix exopolysaccharides of the extracellular polymeric substance (EPS).

Measurement

As the native substrate, cellulose, is a water-insoluble polymer, traditional reducing sugar assays using this substrate can not be employed for the measurement of cellulase activity. Analytical scientists have developed a number of alternative methods.

Viscometric Substrates

A viscometer can be used to measure the decrease in viscosity of a solution containing a water-soluble cellulose derivative such as carboxymethyl cellulose upon incubation with a cellulase sample. The decrease in viscosity is directly proportional to the cellulase activity. While such assays are very sensitive and specific for *endo*-cellulase (*exo*-acting cellulase enzymes produce little or no change in viscosity), they are limited by the fact that it is hard to define activity in conventional enzyme units (micromoles of substrate hydrolyzed or product produced per minute).

Cellooligosaccharide Substrates

The lower DP cello-oligosaccharides (DP2-6) are sufficiently soluble in water to act as viable substrates for cellulase enzymes. However, as these substrates are themselves 'reducing sugars', they are not suitable for use in traditional reducing sugar assays because they generate a high 'blank' value. However their cellulase mediated hydrolysis can be monitored by HPLC or IC methods to gain valuable information on the substrate requirements of a particular cellulase enzyme.

Reduced Cellooligosaccharide Substrates

Cello-oligosaccharides can be chemically reduced through the action of sodium borohydride to produce their corresponding sugar alcohols. These compounds do not react in reducing sugar assays but their hydrolysis products do. This makes borohydride reduced cello-oligosaccharides valuable substrates for the assay of cellulase using traditional reducing sugar assays such as the Nelson-Symogyi method.

Dyed Polysaccharide Substrates

These substrates can be subdivided into two classes-

- Insoluble chromogenic substrates: An insoluble cellulase substrate such as AZ-CL-HE-cellulose absorbs water to create gelatinous particles when placed in solution. This substrate is gradually depolymerised and solubilised by the action of cellulase. The reaction is terminated by adding an alkaline solution to stop enzyme activity and the reaction slurry is filtered or centrifuged. The colour in the filtrate or supernatant is measured and can be related to enzyme activity.

- Soluble chromogenic substrates: A cellulase sample is incubated with a water-soluble substrate such as azo-CM-cellulose, the reaction is terminated and high molecular weight, partially hydrolysed fragments are precipitated from solution with an organic solvent such as ethanol or methoxyethanol. The suspension is mixed thoroughly, centrifuged, and the colour in the supernatant solution (due to small, soluble, dyed fragments) is measured. With the aid of a standard curve, the enzyme activity can be determined.

Enzyme Coupled Reagents

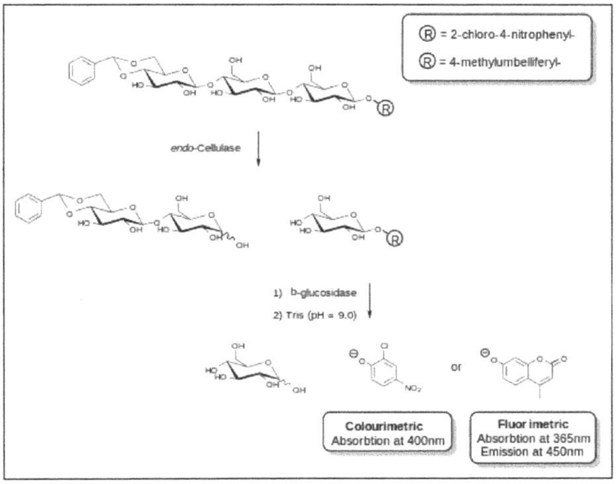

Colourimetric and fluorimetric cellulase substrates can be used in the presence of ancillary β-glucosidase for the specific measurement of *endo*-cellulase activity.

Recently, new reagents have been developed that allow for the specific measurement of *endo*-cellulase. These methods involve the use of functionalised oligosaccharide substrates in the presence of an ancillary enzyme. a cellulase enzyme is able to recognise the trisaccharide fragment of cellulose and cleave this unit. The ancillary enzyme present in the reagent mixture (β-glucosidase) then acts to hydrolyse the fragment containing the chromophore or fluorophore. The assay is terminated by the addition of a basic solution that stops the enzymatic reaction and deprotonates the liberated phenolic compound to produce the phenolate species. The cellulase activity of a given sample is directly proportional to the quantity of phenolate liberated which can be measured using a spectrophotometer. The acetal functionalisation on the non-reducing end of the trisaccharide substrate prevents the action of the ancillary β-glucosidase on the parent substrate.

ABZYMES

Abzymes (from antibody and enzyme) are antibodies that catalyze specific chemical reactions i.e., function as enzymes. Abzymes are also called as catmab (from catalytic monoclonal antibody). Abzymes are usually artificial constructs, but natural abzymes are also known.

The production of the abzymes is based on the following two principles:

1. Enzymes act by binding the transition state of a reactant better than the ground state.

2. Antibodies which bind to specific small molecules can be produced by coupling this small molecule to a protein carrier and using this protein for immunizing experimental animals.

If this small molecule is a transition state analog, then the antibodies that are produced to bind to this molecule will function as enzyme towards the substrate of this reaction. Abzymes are selected from monoclonal antibodies produced by immunizing mice with haptens that mimic the transition state of enzyme catalyzed reactions. For example, 28B4 abzyme catalyzes periodate oxidation of p-nitrotoulene methyl sulphide to sulphoxide, where electrons from the sulfur atom are transferred to the more electronegative oxygen atom.

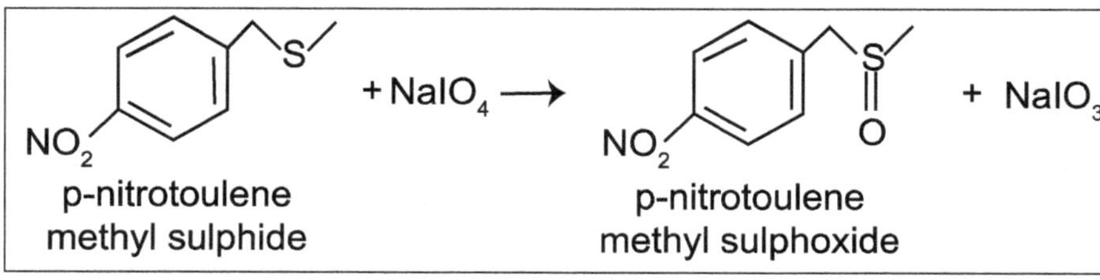

The rate of this reaction is promoted by enzyme catalysts that stabilize the transition state of this reaction, thereby decreasing the activation energy and allowing for more rapid conversion of substrate to product.

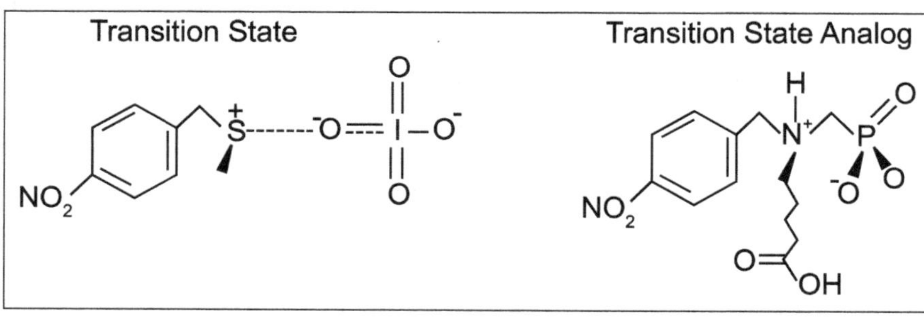

In order to produce abzymes complementary in structure to this transition state, mice were immunized with an aminophosphonic acid hapten as depicted above. Obviously, its structure mirrors the structure and electrostatic properties of the sulphoxide transition state. Of the hapten-binding monoclonal antibodies produced with this hapten, many were found to catalyze sulphide oxidation but with a wide range of binding affinities and catalytic efficiencies. Abzyme 28B4 binds hapten with high affinity and exhibits a corresponding high degree of catalytic efficiency.

Abzymes have proven to be very important tools in Biotechnology and Medical Pathology in combating dreaded human diseases like AIDS, Cancer etc. One of the promising directions in this field consists of the production of abzymes catalyzing rapid cleavage of hazardous compounds, including toxins and drugs such as cocaine.

References

- Katzung BG (2001). Basic and clinical pharmacology:Introduction to autonomic pharmacology (8 ed.). The McGraw Hill Companies. pp. 75–91. ISBN 978-0-07-160405-5
- Proteolytic-enzyme, science: britannica.com, Retrieved 15 April, 2019
- Nichols JD, Xiang S, Schindelin H, Rajagopalan KV (Jan 2007). "Mutational analysis of Escherichia coli MoeA: two functional activities map to the active site cleft". Biochemistry. 46 (1): 78–86. doi:10.1021/bi061551q. PMC 1868504. PMID 17198377
- Study-notes-on-abzymes-with-diagram, enzymes-22960: biologydiscussion.com, Retrieved 25 June, 2019
- Andrews, Lary. "Supplemental Enzymes for Digestion". Health and Healing Research. Archived from the original on 27 July 2013. Retrieved 9 December, 2013

Enzyme Kinetics

CHAPTER 3

The branch of science which deals with the study of the chemical reactions which are catalyzed by enzymes is known as enzyme kinetics. The increase which happens due to the presence of an enzyme in the rate of a process is known as enzyme catalysis. The chapter closely examines these key concepts of enzyme kinetics to provide an extensive understanding of the subject.

Enzyme kinetics involves the measurement of the rate at which chemical reactions that are catalyzed by enzymes occur. Knowledge about the kinetics of an enzyme can reveal useful information about its catalytic mechanism, role in metabolism, factors that impact its activity, and mechanisms of inhibition.

Rate of Reaction

Enzymes are thought to form a complex with the substrates to catalyze the reaction. This process can be illustrated with the simplified equation, where e is the enzyme, S is the substrate, and P is the product:

$$E + S \Leftrightarrow ES \Rightarrow + P + E$$

The first step of the equation, which is reversible, has the reaction rate constant of k_{+1} to produce the enzyme substrate complex and k_{-1} for the reverse reaction. The reaction rate constant for the second step of the equation, which is not reversible, is k_{+2}.

The rate of reaction (v), which is the rate at which the product is formed, is defined by the following equation:

$$V = d[P] / dt = k_{+2}[ES]$$

The square brackets in the above equation represent the molar concentration of the substance specified within, so [P] refers to the molar concentration of the product and [ES] the molar concentration of the enzyme substrate.

V_{max} and K_m

Leonor Michaelis and Maud Menten introduced a mathematical illustration to describe the action of enzymes with two constants, V_{max} and K_m.

The maximal velocity (V_{max}) refers to the point at which the increase the concentration of the substrate does not increase the rate of a reaction catalyzed by an enzyme.

This occurs because the substrate molecules saturate the active sites of the enzyme and are not able to form more complexes with the enzyme. This value is given as a rate (mmol/s), which is the maximum velocity of the reaction when the enzyme is saturated.

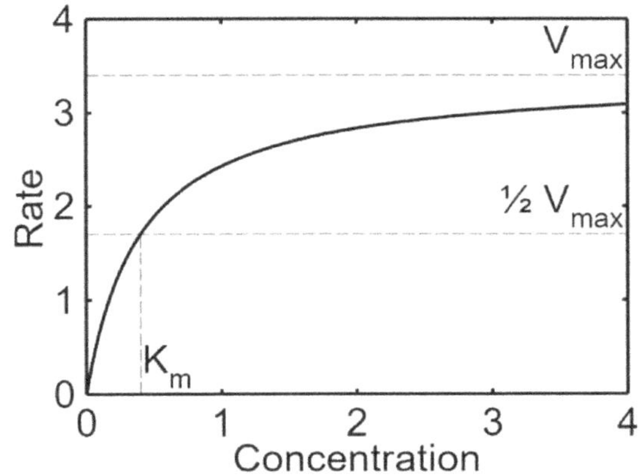

Michaelis–Menten saturation curve of an enzyme reaction.
Parameter values used are Vmax=3.4 and Km=1.7.

The Michaelis constant (K_m) is the concentration of the substrate when half of the active binding sites of an enzyme are occupied by the substrate. The constant helps to depict the affinity of the enzyme for their substrate. This value is given as the concentration of the substrate (mM) at half of V_{max}. An enzyme with a high K_m has a low affinity for the substrate, and a high concentration of the substrate is needed in order for the enzyme to become saturated. Conversely, an enzyme with a high K_m has a high affinity for the substrate and the enzyme may become saturated even with a low amount of substrate.

Applications of Enzyme Kinetics

There are many practical uses of enzyme kinetics. For example, the kinetic constants can help explain how enzymes work and assist in the prediction of the behavior of enzymes in living organisms.

V_{max} and K_m both play a key role in understanding the metabolism of the human body. Knowledge of the enzyme kinetic constants allows us to gain a better understanding of the enzymes and processes that take place in human metabolism. This knowledge could then be used for medical purposes to improve patient health outcomes.

ENZYME CATALYSIS

Enzyme catalysis is the increase in the rate of a process by a biological molecule, an "enzyme". Most enzymes are proteins, and most such processes are chemical reactions. Within the enzyme, generally catalysis occurs at a localized site, called the active site.

Most enzymes are made predominantly of proteins, either a single protein chain or many such chains in a multi-subunit complex. Enzymes often also incorporate non-protein components, such as metal ions or specialized organic molecules known as cofactor (e.g. adenosine triphosphate). Many cofactors are vitamins, and their role as vitamins is directly linked to their use in the catalysis of biological process within metabolism. Catalysis of biochemical reactions in the cell is vital since many but not all metabolically essential reactions have very low rates when uncatalysed. One driver of protein evolution is the optimization of such catalytic activities, although only the most crucial enzymes operate near catalytic efficiency limits, and many enzymes are far from optimal. Important factors in enzyme catalysis include general acid and base catalysis, orbital steering, entropic restriction, orientation effects (i.e. lock and key catalysis), as well as motional effects involving protein dynamics.

Mechanisms of enzyme catalysis vary, but are all similar in principle to other types of chemical catalysis in that the crucial factor is a reduction of energy barrier(s) separating the reactants from the products. The reduction of activation energy (E_a) increases the fraction of reactant molecules that can overcome this barrier and form the product. An important principle is that that since they only reduce energy barriers between products and reactants, enzymes always catalyze reactions in both directions, and cannot drive a reaction forward or affect the equilibrium position - only the speed with which is it achieved. As with other catalysts, the enzyme is not consumed or changed by the reaction (as a substrate is) but is recycled such that a single enzyme performs many rounds of catalysis.

Induced Fit

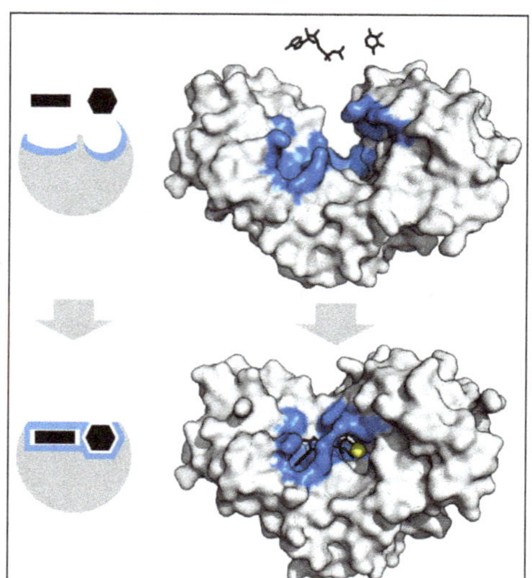

Enzyme changes shape by induced fit upon substrate binding to form enzyme-substrate complex. Hexokinase has a large induced fit motion that closes over the substrates adenosine triphosphate and xylose. Binding sites in blue, substrates in black and Mg^{2+} cofactor in yellow.

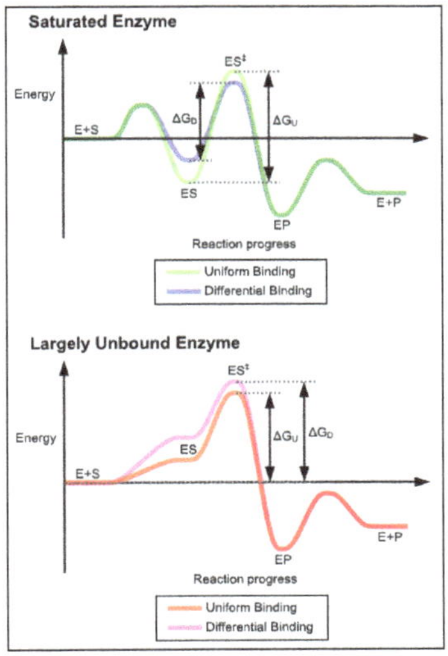

The different mechanisms of substrate binding.

The favored model for the enzyme-substrate interaction is the induced fit model. This model proposes that the initial interaction between enzyme and substrate is relatively weak, but that these weak interactions rapidly induce conformational changes in the enzyme that strengthen binding.

The advantages of the induced fit mechanism arise due to the stabilizing effect of strong enzyme binding. There are two different mechanisms of substrate binding: uniform binding, which has strong substrate binding, and differential binding, which has strong transition state binding. The stabilizing effect of uniform binding increases both substrate and transition state binding affinity, while differential binding increases only transition state binding affinity. Both are used by enzymes and have been evolutionarily chosen to minimize the activation energy of the reaction. Enzymes that are saturated, that is, have a high affinity substrate binding, require differential binding to reduce the energy of activation, whereas small substrate unbound enzymes may use either differential or uniform binding.

These effects have led to most proteins using the differential binding mechanism to reduce the energy of activation, so most substrates have high affinity for the enzyme while in the transition state. Differential binding is carried out by the induced fit mechanism - the substrate first binds weakly, then the enzyme changes conformation increasing the affinity to the transition state and stabilizing it, so reducing the activation energy to reach it.

It is important to clarify, however, that the induced fit concept cannot be used to rationalize catalysis. That is, the chemical catalysis is defined as the reduction of $E_a^\ddagger$ (when the system is already in the $ES^\ddagger$) relative to $E_a^\ddagger$ in the uncatalyzed reaction in water (without the enzyme). The induced fit only suggests that the barrier is lower in the closed form of the enzyme but does not tell us what the reason for the barrier reduction is.

Induced fit may be beneficial to the fidelity of molecular recognition in the presence of competition and noise via the conformational proofreading mechanism.

Mechanisms of an Alternative Reaction Route

These conformational changes also bring catalytic residues in the active site close to the chemical bonds in the substrate that will be altered in the reaction. After binding takes place, one or more mechanisms of catalysis lowers the energy of the reaction's transition state, by providing an alternative chemical pathway for the reaction. There are six possible mechanisms of "over the barrier" catalysis as well as a "through the barrier" mechanism.

Proximity and Orientation

Enzyme-substrate interactions align the reactive chemical groups and hold them close together in an optimal geometry, which increases the rate of the reaction. This reduces the entropy of the reactants and thus makes addition or transfer reactions less unfavorable, since a reduction in the overall entropy when two reactants become a single product. However this is a general effect and is seen in non-addition or transfer reactions where it occurs due to an increase in the "effective concentration" of the reagents. This is understood when considering how increases in concentration leads to increases in reaction rate: essentially when the reactants are more concentrated, they collide more often and so react more often. In enzyme catalysis, the binding of the reagents to the enzyme restricts the conformational space of the reactants, holding them in the 'proper orientation' and close to each other, so that the collide more frequently, and with the correct geometry, to facilitate the desired reaction. The "effective concentration" is the concentration the reactant would have to be, free in solution, to experiences the same collisional frequency. Often such theoretical effective concentrations are unphysical and impossible to realize in reality - which is a testament to the great catalytic power of many enzymes, with massive rate increases over the uncatalyzed state.

For example:

Similar reactions will occur far faster if the reaction is intramolecular.

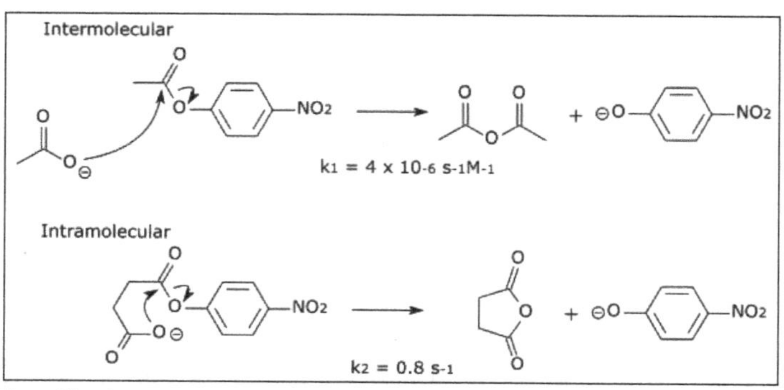

The effective concentration of acetate in the intramolecular reaction can be estimated as $k_2/k_1 = 2 \times 10^5$ Molar.

However, the situation might be more complex, since modern computational studies have established that traditional examples of proximity effects cannot be related directly to enzyme entropic effects. Also, the original entropic proposal has been found to largely overestimate the contribution of orientation entropy to catalysis.

Proton Donors or Acceptors

Proton donors and acceptors, i.e. acids and base may donate and accept protons in order to stabilize developing charges in the transition state. This is related to the overall principle of catalysis, that of reducing energy barriers, since in general transition stated are high energy states, and by stabilizing them this high energy is reduced, lowering the barrier. A key feature of enzyme catalysis over many non-biological catalysis, is that both acid and base catalysis can be combined in the same reaction. In many abiotic systems, acids (large [H+]) or bases (large concentration H+ sinks, or species with electron pairs) can increase the rate of the reaction; but of course the environamen can only have one overall pH (measure of acidity or basicity (alkalinity)). However since enzymes are large molecules, they can position both acid groups and basic groups in their active site to interact with their substrates, an meploy both modes independed of the bulk pH.

Often general acid of base catalysis is employed to activate nucleophile and/or electrophile groups, or to stabilizing leaving groups. Many amino acids with acidic or basic groups are this employed in the active site, such as the glutamic and aspartic acid, histidine, cystine, tyrosine, lysine and arginine, as well as serine and threonine. In addition the peptide backbone, with carbonyl and amide N groups is often employed. Cystine and Histidine are very commonly involved, since they both have a pKa close to neutral pH and can therefore both accept and donate protons.

Many reaction mechanisms involving acid/base catalysis assume a substantially altered pKa. This alteration of pKa is possible through the local environment of the residue.

Conditions	Acids	Bases
Hydrophobic environment	Increase pKa	Decrease pKa
Adjacent residues of like charge	Increase pKa	Decrease pKa
Salt bridge (and hydrogen bond) formation	Decrease pKa	Increase pKa

pKa can also be influenced significantly by the surrounding environment, to the extent that residues which are basic in solution may act as proton donors, and vice versa.

It is important to clarify that the modification of the pKa's is a pure part of the electrostatic mechanism. Furthermore, the catalytic effect of the above example is mainly associated with the reduction of the pKa of the oxyanion and the increase in the pKa of the histidine, while the proton transfer from the serine to the histidine is not catalyzed significantly, since it is not the rate determining barrier.

For example:

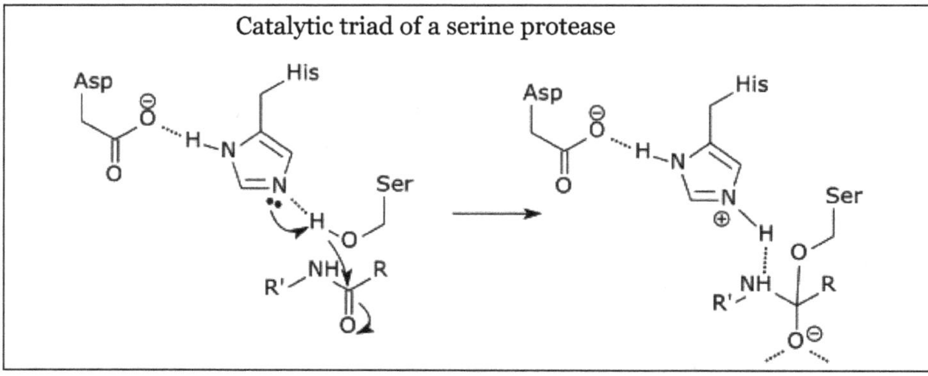

The initial step of the serine protease catalytic mechanism involves the histidine of the active site accepting a proton from the serine residue. This prepares the serine as a nucleophile to attack the amide bond of the substrate. This mechanism includes donation of a proton from serine (a base, pKa 14) to histidine (an acid, pKa 6), made possible due to the local environment of the bases.

Electrostatic Catalysis

Stabilization of charged transition states can also be by residues in the active site forming ionic bonds (or partial ionic charge interactions) with the intermediate. These bonds can either come from acidic or basic side chains found on amino acids such as lysine, arginine, aspartic acid or glutamic acid or come from metal cofactors such as zinc. Metal ions are particularly effective and can reduce the pKa of water enough to make it an effective nucleophile.

Systematic computer simulation studies established that electrostatic effects give, by far, the largest contribution to catalysis. In particular, it has been found that enzyme provides an environment which is more polar than water, and that the ionic transition states are stabilized by fixed dipoles. This is very different from transition state stabilization in water, where the water molecules must pay with “reorganization energy”. In order to stabilize ionic and charged states. Thus, the catalysis is associated with the fact that the enzyme polar groups are preorganized

The magnitude of the electrostatic field exerted by an enzyme’s active site has been shown to be highly correlated with the enzyme’s catalytic rate enhancement

Binding of substrate usually excludes water from the active site, thereby lowering the local dielectric constant to that of an organic solvent. This strengthens the electrostatic interactions between the charged/polar substrates and the active sites. In addition, studies have shown that the charge distributions about the active sites are arranged so as to stabilize the transition states of the catalyzed reactions. In several enzymes, these charge distributions apparently serve to guide polar substrates toward their binding sites so that the rates of these enzymatic reactions are greater than their apparent diffusion-controlled limits.

For example:

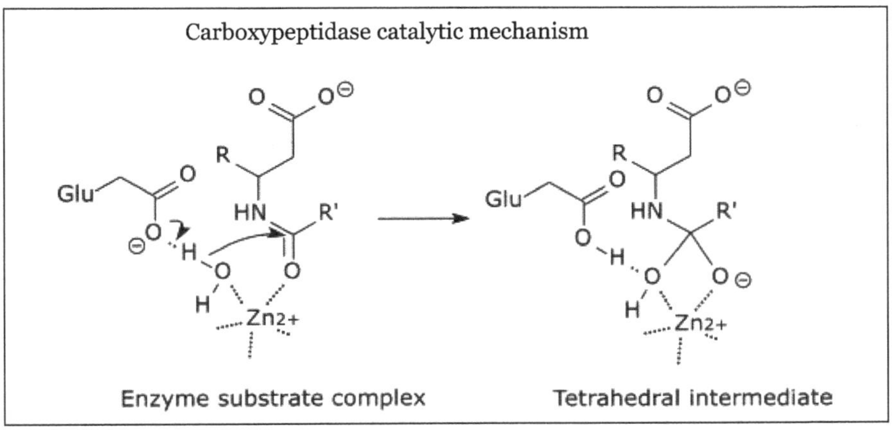

The tetrahedral intermediate is stabilised by a partial ionic bond between the Zn^{2+} ion and the negative charge on the oxygen.

Covalent Catalysis

Covalent catalysis involves the substrate forming a transient covalent bond with residues in the enzyme active site or with a cofactor. This adds an additional covalent intermediate to the reaction, and helps to reduce the energy of later transition states of the reaction. The covalent bond must, at a later stage in the reaction, be broken to regenerate the enzyme. This mechanism is utilised by the catalytic triad of enzymes such as proteases like chymotrypsin and trypsin, where an acyl-enzyme intermediate is formed. An alternative mechanism is schiff base formation using the free amine from a lysine residue, as seen in the enzyme aldolase during glycolysis.

Some enzymes utilize non-amino acid cofactors such as pyridoxal phosphate (PLP) or thiamine pyrophosphate (TPP) to form covalent intermediates with reactant molecules. Such covalent intermediates function to reduce the energy of later transition states, similar to how covalent intermediates formed with active site amino acid residues allow stabilization, but the capabilities of cofactors allow enzymes to carryout reactions that amino acid side residues alone could not. Enzymes utilizing such cofactors include the PLP-dependent enzyme aspartate transaminase and the TPP-dependent enzyme pyruvate dehydrogenase.

Rather than lowering the activation energy for a reaction pathway, covalent catalysis provides an alternative pathway for the reaction (via to the covalent intermediate) and so is distinct from true catalysis. For example, the energetics of the covalent bond to the serine molecule in chymotrypsin should be compared to the well-understood covalent bond to the nucleophile in the uncatalyzed solution reaction. A true proposal of a covalent catalysis (where the barrier is lower than the corresponding barrier in solution) would require, for example, a partial covalent bond to the transition state by an enzyme group (e.g., a very strong hydrogen bond), and such effects do not contribute significantly to catalysis.

Metal ion Catalysis

A metal ion in the active site participates in catalysis by coordinating charge stabilization and shielding. Because of a metal's positive charge, only negative charges can be stabilized through metal ions. However, metal ions are advantageous in biological catalysis because they are not affected by changes in pH. Metal ions can also act to ionize water by acting as a Lewis acid. Metal ions may also be agents of oxidation and reduction.

Bond Strain

This is the principal effect of induced fit binding, where the affinity of the enzyme to the transition state is greater than to the substrate itself. This induces structural rearrangements which strain substrate bonds into a position closer to the conformation of the transition state, so lowering the energy difference between the substrate and transition state and helping catalyze the reaction.

However, the strain effect is, in fact, a ground state destabilization effect, rather than transition state stabilization effect. Furthermore, enzymes are very flexible and they cannot apply large strain effect.

In addition to bond strain in the substrate, bond strain may also be induced within the enzyme itself to activate residues in the active site.

For example:

Substrate, bound substrate, and transition state conformations of lysozyme.

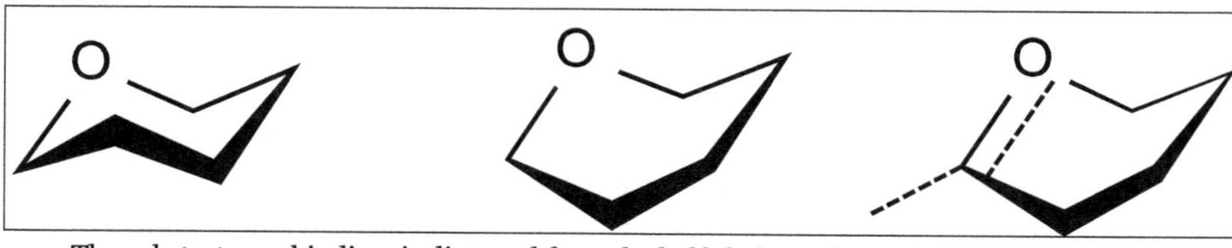

The substrate, on binding, is distorted from the half chair conformation of the hexose ring (because of the steric hindrance with amino acids of the protein forcing the equatorial c6 to be in the axial position) into the chair conformation.

Quantum Tunneling

These traditional "over the barrier" mechanisms have been challenged in some cases by models and observations of "through the barrier" mechanisms (quantum tunneling).

Some enzymes operate with kinetics which are faster than what would be predicted by the classical $\Delta G^{\ddagger}$. In "through the barrier" models, a proton or an electron can tunnel through activation barriers. Quantum tunneling for protons has been observed in tryptamine oxidation by aromatic amine dehydrogenase.

Quantum tunneling does not appear to provide a major catalytic advantage, since the tunneling contributions are similar in the catalyzed and the uncatalyzed reactions in solution. However, the tunneling contribution (typically enhancing rate constants by a factor of ~1000 compared to the rate of reaction for the classical 'over the barrier' route) is likely crucial to the viability of biological organisms. This emphasizes the general importance of tunneling reactions in biology. In 1971-1972 the first quantum-mechanical model of enzyme catalysis was formulated.

Active Enzyme

The binding energy of the enzyme-substrate complex cannot be considered as an external energy which is necessary for the substrate activation. The enzyme of high energy content may firstly transfer some specific energetic group X_1 from catalytic site of the enzyme to the final place of the first bound reactant, then another group X_2 from the second bound reactant (or from the second group of the single reactant) must be transferred to active site to finish substrate conversion to product and enzyme regeneration.

We can present the whole enzymatic reaction as a two coupling reactions:

$$S_1 + EX_1 \rightarrow S_1EX_1 \rightarrow P_1 + EP_2$$

$$S_2 + EP_2 \rightarrow S_2EP_2 \rightarrow P_2 + EX_2$$

It may be seen from reaction $S_1 + EX_1 \rightarrow S_1EX_1 \rightarrow P_1 + EP_2$ that the group X_1 of the active enzyme appears in the product due to possibility of the exchange reaction inside enzyme to avoid both electrostatic inhibition and repulsion of atoms. So we represent the active enzyme as a powerful reactant of the enzymatic reaction. The reaction $S_2 + EP_2 \rightarrow S_2EP_2 \rightarrow P_2 + EX_2$ shows incomplete conversion of the substrate because its group X_2 remains inside enzyme. This approach as idea had formerly proposed relying on the hypothetical extremely high enzymatic conversions (catalytically perfect enzyme).

The crucial point for the verification of the present approach is that the catalyst must be a complex of the enzyme with the transfer group of the reaction. This chemical aspect is supported by the well-studied mechanisms of the several enzymatic reactions. Consider the reaction of peptide bond hydrolysis catalyzed by a pure protein α-chymotrypsin (an enzyme acting without a cofactor), which is a well-studied member of the serine proteases family.

We present the experimental results for this reaction as two chemical steps:

$$S_1 + EH \rightarrow P_1 + EP_2$$

$$EP_2 + H - O - H \rightarrow EH + P_2$$

where S_1 is a polypeptide, P_1 and P_2 are products. The first chemical step $S_1 + EH \rightarrow P_1 + EP_2$ includes the formation of a covalent acyl-enzyme intermediate. The second step $EP_2 + H - O - H \rightarrow EH + P_2$ is the deacylation step. It is important to note that the group H+, initially found on the enzyme, but not in water, appears in the product before the step of hydrolysis, therefore it may be considered as an additional group of the enzymatic reaction.

Thus, the reaction $S_1 + EH \rightarrow P_1 + EP_2$ shows that the enzyme acts as a powerful reactant of the reaction. According to the proposed concept, the H transport from the enzyme promotes the first reactant conversion, breakdown of the first initial chemical bond (between groups P_1 and P_2). The step of hydrolysis leads to a breakdown of the second chemical bond and regeneration of the enzyme.

The proposed chemical mechanism does not depend on the concentration of the substrates or products in the medium. However, a shift in their concentration mainly causes free energy changes in the first and final steps of the reactions $S_1 + EX_1 \rightarrow S_1EX_1 \rightarrow P_1 + EP_2$ and $S_2 + EP_2 \rightarrow S_2EP_2 \rightarrow P_2 + EX_2$ due to the changes in the free energy content of every molecule, whether S or P, in water solution. This approach is in accordance with the following mechanism of muscle contraction. The final step of ATP hydrolysis in skeletal muscle is the product release caused by the association of myosin heads with actin. The closing of the actin-binding cleft during the association reaction is structurally coupled with the opening of the nucleotide-binding pocket on the myosin active site.

Notably, the final steps of ATP hydrolysis include the fast release of phosphate and the slow release of ADP. The release of a phosphate anion from bound ADP anion into water solution may be considered as an exergonic reaction because the phosphate anion has low molecular mass.

Thus, we arrive at the conclusion that the primary release of the inorganic phosphate $H_2PO_4^-$ leads to transformation of a significant part of the free energy of ATP hydrolysis into the kinetic energy of the solvated phosphate, producing active streaming. This assumption of a local mechano-chemical transduction is in accord with Tirosh's mechanism of muscle contraction, where the muscle force derives from an integrated action of active streaming created by ATP hydrolysis.

Examples of Catalytic Mechanisms

In reality, most enzyme mechanisms involve a combination of several different types of catalysis.

Triose Phosphate Isomerase

Triose phosphate isomerase catalyses the reversible interconvertion of the two triose phosphates isomers dihydroxyacetone phosphate and D-glyceraldehyde 3-phosphate.

Trypsin

Trypsin is a serine protease that cleaves protein substrates after lysine or arginine residues using a catalytic triad to perform covalent catalysis, and an oxyanion hole to stabilise charge-buildup on the transition states.

Aldolase

Aldolase catalyses the breakdown of fructose 1,6-bisphosphate (F-1,6-BP) into glyceraldehyde 3-phosphate and dihydroxyacetone phosphate (DHAP).

Enzyme Diffusivity

The advent of single-molecule studies led in the 2010s to the observation that the movement of untethered enzymes increases with increasing substrate concentration and increasing reaction enthalpy. Subsequent observations suggest that this increase in diffusivity is driven by transient displacement of the enzyme's center of mass, resulting in a "recoil effect that propels the enzyme".

ENZYME ASSAYS

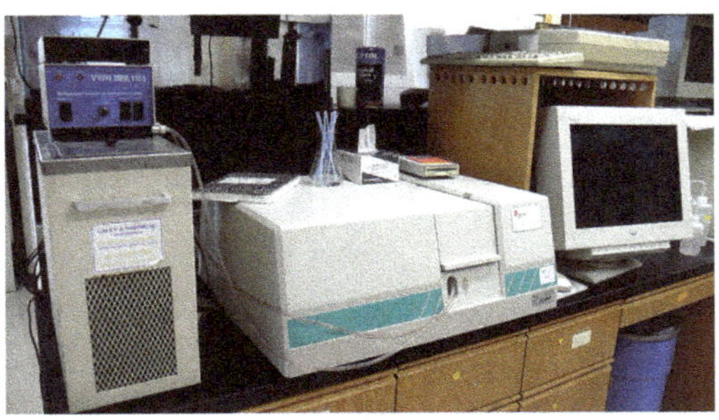

Beckman DU640 UV/Vis spectrophotometer.

Enzyme assays are laboratory methods for measuring enzymatic activity. They are vital for the study of enzyme kinetics and enzyme inhibition.

Enzyme Units

The quantity or concentration of an enzyme can be expressed in molar amounts, as with any other chemical, or in terms of activity in enzyme units.

Enzyme Activity

Enzyme activity = moles of substrate converted per unit time = rate × reaction volume. Enzyme activity is a measure of the quantity of active enzyme present and is thus dependent on conditions, *which should be specified*. The SI unit is the katal, 1 katal = 1 mol s^{-1}, but this is an excessively large unit. A more practical and commonly used value is enzyme unit (U) = 1 μmol min^{-1}. 1 U corresponds to 16.67 nanokatals.

Enzyme activity as given in katal generally refers to that of the assumed natural target substrate of the enzyme. Enzyme activity can also be given as that of certain standardized substrates, such as gelatin, then measured in *gelatin digesting units* (GDU), or milk proteins, then measured in *milk clotting units* (MCU). The units GDU and MCU are based on how fast one gram of the enzyme will digest gelatin or milk proteins, respectively. 1 GDU equals approximately 1.5 MCU.

An increased amount of substrate will increase the rate of reaction with enzymes, however once past a certain point, the rate of reaction will level out because the amount of active sites available has stayed constant.

Specific Activity

The specific activity of an enzyme is another common unit. This is the activity of an enzyme per milligram of total protein (expressed in μmol min^{-1} mg^{-1}). Specific activity gives a measurement of enzyme purity in the mixture. It is the micro moles of product formed by an enzyme in a given amount of time (minutes) under given conditions per milligram of total proteins. Specific activity is equal to the rate of reaction multiplied by the volume of reaction divided by the mass of total protein. The SI unit is katal/kg, but a more practical unit is μmol/mgmin.

Specific activity is a measure of *enzyme processivity* (the capability of enzyme to be processed), at a specific (usually saturating) substrate concentration, and is usually constant for a pure enzyme.

An active site titration process can be done for the elimination of errors arising from differences in cultivation batches and/or misfolded enzyme and similar issues. This is a measure of the amount of active enzyme, calculated by e.g. titrating the amount of active sites present by employing an irreversible inhibitor. The specific activity should then be expressed as μmol min^{-1} mg^{-1} active enzyme. If the molecular weight of the enzyme is known, the turnover number, or μmol product per second per μmol of active enzyme, can be calculated from the specific activity. The turnover number can be visualized as the number of times each enzyme molecule carries out its catalytic cycle per second.

The rate of a reaction is the concentration of substrate disappearing (or product produced) per unit time (mol L^{-1} s^{-1}).

The % purity is 100% × (specific activity of enzyme sample / specific activity of pure enzyme). The impure sample has lower specific activity because some of the mass is not actually enzyme. If the specific activity of 100% pure enzyme is known, then an impure sample will have a lower specific activity, allowing purity to be calculated.

Types of Assay

All enzyme assays measure either the consumption of substrate or production of product over time. A large number of different methods of measuring the concentrations of substrates and products exist and many enzymes can be assayed in several different ways. Biochemists usually study enzyme-catalysed reactions using four types of experiments:

- Initial rate experiments: When an enzyme is mixed with a large excess of the substrate, the enzyme-substrate intermediate builds up in a fast initial transient. Then the reaction achieves a steady-state kinetics in which enzyme substrate intermediates remains approximately constant over time and the reaction rate changes relatively slowly. Rates are measured for a short period after the attainment of the quasi-steady state, typically by monitoring the accumulation of product with time. Because the measurements are carried out for a very short period and because of the large excess of substrate, the approximation that the amount of free substrate is approximately equal to the amount of the initial substrate can be made. The initial rate experiment is the simplest to perform and analyze, being relatively free from complications such as back-reaction and enzyme degradation. It is therefore by far the most commonly used type of experiment in enzyme kinetics.

- Progress curve experiments: In these experiments, the kinetic parameters are determined from expressions for the species concentrations as a function of time. The concentration of the substrate or product is recorded in time after the initial fast transient and for a sufficiently long period to allow the reaction to approach equilibrium. Progress curve experiments were widely used in the early period of enzyme kinetics, but are less common now.

- Transient kinetics experiments: In these experiments, reaction behaviour is tracked during the initial fast transient as the intermediate reaches the steady-state kinetics period. These experiments are more difficult to perform than either of the above two classes because they require specialist techniques (such as flash photolysis of caged compounds) or rapid mixing (such as stopped-flow, quenched flow or continuous flow).

- Relaxation experiments: In these experiments, an equilibrium mixture of enzyme, substrate and product is perturbed, for instance by a temperature, pressure or pH jump, and the return to equilibrium is monitored. The analysis of these experiments requires consideration of the fully reversible reaction. Moreover, relaxation experiments are relatively insensitive to mechanistic details and are thus not typically used for mechanism identification, although they can be under appropriate conditions.

Enzyme assays can be split into two groups according to their sampling method: continuous assays, where the assay gives a continuous reading of activity, and discontinuous assays, where samples are taken, the reaction stopped and then the concentration of substrates/products determined.

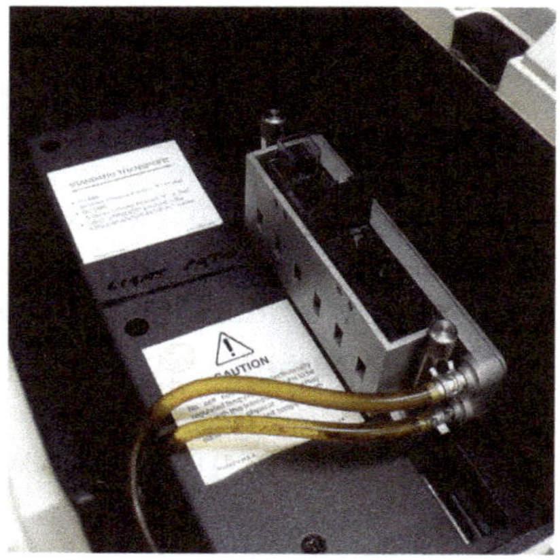

Temperature-controlled cuvette holder in a spectrophotometer.

Continuous Assays

Continuous assays are most convenient, with one assay giving the rate of reaction with no further work necessary. There are many different types of continuous assays.

Spectrophotometric

In spectrophotometric assays, you follow the course of the reaction by measuring a change in how much light the assay solution absorbs. If this light is in the visible region you can actually see a change in the color of the assay, and these are called colorimetric assays. The MTT assay, a redox assay using a tetrazolium dye as substrate is an example of a colorimetric assay.

UV light is often used, since the common coenzymes NADH and NADPH absorb UV light in their reduced forms, but do not in their oxidized forms. An oxidoreductase using NADH as a substrate could therefore be assayed by following the decrease in UV absorbance at a wavelength of 340 nm as it consumes the coenzyme.

Direct Versus Coupled Assays

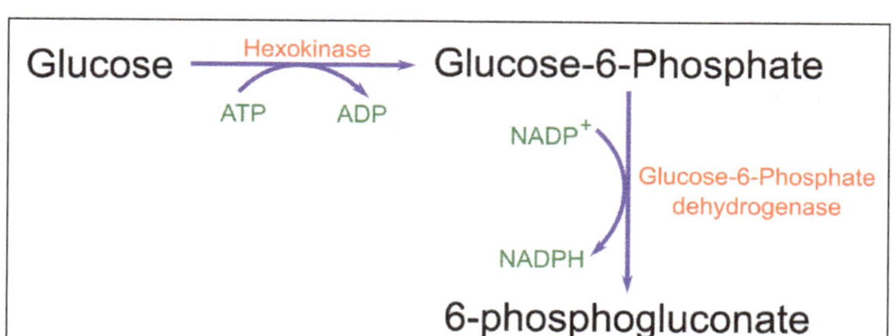

Coupled assay for hexokinase using glucose-6-phosphate dehydrogenase.

Even when the enzyme reaction does not result in a change in the absorbance of light, it can still be possible to use a spectrophotometric assay for the enzyme by using a coupled assay. Here, the product of one reaction is used as the substrate of another, easily detectable reaction. For example, the coupled assay for the enzyme hexokinase, which can be assayed by coupling its production of glucose-6-phosphate to NADPH production, using glucose-6-phosphate dehydrogenase.

Fluorometric

Fluorescence is when a molecule emits light of one wavelength after absorbing light of a different wavelength. Fluorometric assays use a difference in the fluorescence of substrate from product to measure the enzyme reaction. These assays are in general much more sensitive than spectrophotometric assays, but can suffer from interference caused by impurities and the instability of many fluorescent compounds when exposed to light.

An example of these assays is again the use of the nucleotide coenzymes NADH and NADPH. Here, the reduced forms are fluorescent and the oxidised forms non-fluorescent. Oxidation reactions can therefore be followed by a decrease in fluorescence and reduction reactions by an increase. Synthetic substrates that release a fluorescent dye in an enzyme-catalyzed reaction are also available, such as 4-methylumbelliferyl-β-D-galactoside for assaying β-galactosidase.

Calorimetric

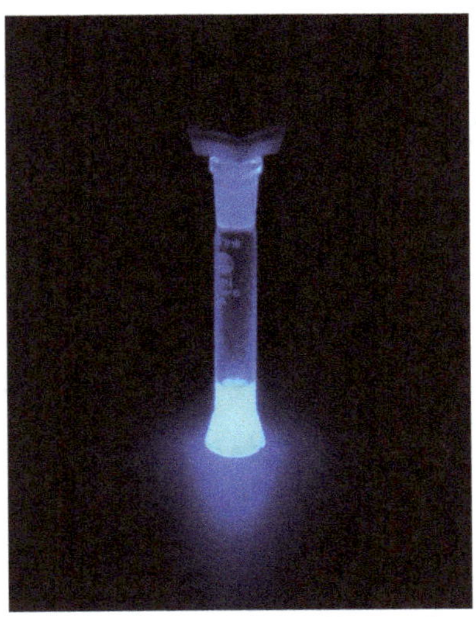

Chemiluminescence of luminol.

Calorimetry is the measurement of the heat released or absorbed by chemical reactions. These assays are very general, since many reactions involve some change in heat and with use of a microcalorimeter, not much enzyme or substrate is required. These assays can be used to measure reactions that are impossible to assay in any other way.

Chemiluminescent

Chemiluminescence is the emission of light by a chemical reaction. Some enzyme reactions produce light and this can be measured to detect product formation. These types of assay can be extremely sensitive, since the light produced can be captured by photographic film over days or weeks, but can be hard to quantify, because not all the light released by a reaction will be detected.

The detection of horseradish peroxidase by enzymatic chemiluminescence (ECL) is a common method of detecting antibodies in western blotting. Another example is the enzyme luciferase, this is found in fireflies and naturally produces light from its substrate luciferin.

Light Scattering

Static light scattering measures the product of weight-averaged molar mass and concentration of macromolecules in solution. Given a fixed total concentration of one or more species over the measurement time, the scattering signal is a direct measure of the weight-averaged molar mass of the solution, which will vary as complexes form or dissociate. Hence the measurement quantifies the stoichiometry of the complexes as well as kinetics. Light scattering assays of protein kinetics is a very general technique that does not require an enzyme.

Microscale Thermophoresis

Microscale thermophoresis (MST) measures the size, charge and hydration entropy of molecules/substrates at equilibrium. The thermophoretic movement of a fluorescently labeled substrate changes significantly as it is modified by an enzyme. This enzymatic activity can be measured with high time resolution in real time. The material consumption of the all optical MST method is very low, only 5 μl sample volume and 10nM enzyme concentration are needed to measure the enzymatic rate constants for activity and inhibition. MST allows analysts to measure the modification of two different substrates at once (multiplexing) if both substrates are labeled with different fluorophores. Thus substrate competition experiments can be performed.

Discontinuous Assays

Discontinuous assays are when samples are taken from an enzyme reaction at intervals and the amount of product production or substrate consumption is measured in these samples.

Radiometric

Radiometric assays measure the incorporation of radioactivity into substrates or its release from substrates. The radioactive isotopes most frequently used in these assays are

^{14}C, ^{32}P, ^{35}S and ^{125}I. Since radioactive isotopes can allow the specific labelling of a single atom of a substrate, these assays are both extremely sensitive and specific. They are frequently used in biochemistry and are often the only way of measuring a specific reaction in crude extracts (the complex mixtures of enzymes produced when you lyse cells).

Radioactivity is usually measured in these procedures using a scintillation counter.

Chromatographic

Chromatographic assays measure product formation by separating the reaction mixture into its components by chromatography. This is usually done by high-performance liquid chromatography (HPLC), but can also use the simpler technique of thin layer chromatography. Although this approach can need a lot of material, its sensitivity can be increased by labelling the substrates/products with a radioactive or fluorescent tag. Assay sensitivity has also been increased by switching protocols to improved chromatographic instruments (e.g. ultra-high pressure liquid chromatography) that operate at pump pressure a few-fold higher than HPLC instruments.

Factors to Control in Assays

A pressure chamber for measuring enzyme activity at high pressure.

- Salt Concentration: Most enzymes cannot tolerate extremely high salt concentrations. The ions interfere with the weak ionic bonds of proteins. Typical enzymes are active in salt concentrations of 1-500 mM. As usual there are exceptions such as the halophilic algae and bacteria.
- Effects of Temperature: All enzymes work within a range of temperature specific to the organism. Increases in temperature generally lead to increases in reaction rates. There is a limit to the increase because higher temperatures lead to a sharp decrease in reaction rates. This is due to the denaturating (alteration) of protein structure resulting from the breakdown of the weak ionic and hydrogen bonding that stabilize the three-dimensional structure of the enzyme active site. The "optimum" temperature for human enzymes is usually between 35 and

40 °C. The average temperature for humans is 37 °C. Human enzymes start to denature quickly at temperatures above 40 °C. Enzymes from thermophilic archaea found in the hot springs are stable up to 100 °C. However, the idea of an "optimum" rate of an enzyme reaction is misleading, as the rate observed at any temperature is the product of two rates, the reaction rate and the denaturation rate. If you were to use an assay measuring activity for one second, it would give high activity at high temperatures, however if you were to use an assay measuring product formation over an hour, it would give you low activity at these temperatures.

- Effects of pH: Most enzymes are sensitive to pH and have specific ranges of activity. All have an optimum pH. The pH can stop enzyme activity by denaturating (altering) the three-dimensional shape of the enzyme by breaking ionic, and hydrogen bonds. Most enzymes function between a pH of 6 and 8; however pepsin in the stomach works best at a pH of 2 and trypsin at a pH of 8.
- Substrate Saturation: Increasing the substrate concentration increases the rate of reaction (enzyme activity). However, enzyme saturation limits reaction rates. An enzyme is saturated when the active sites of all the molecules are occupied most of the time. At the saturation point, the reaction will not speed up, no matter how much additional substrate is added. The graph of the reaction rate will plateau.
- Level of crowding, large amounts of macromolecules in a solution will alter the rates and equilibrium constants of enzyme reactions, through an effect called macromolecular crowding.

MICHAELIS–MENTEN KINETICS

Michaelis–Menten kinetics is one of the best-known models of enzyme kinetics. It is named after German biochemist Leonor Michaelis and Canadian physician Maud Menten. The model takes the form of an equation describing the rate of enzymatic reactions, by relating reaction rate v (rate of formation of product,) $[\mathrm{P}]$ to $[\mathrm{S}]$, the concentration of a substrate S. Its formula is given by,

$$v = \frac{\mathrm{d}[P]}{\mathrm{d}t} = \frac{V_{\max}[S]}{K_{\mathrm{M}} + [S]}.$$

This equation is called the Michaelis–Menten equation. Here, $V_{\max}$ represents the maximum rate achieved by the system(which happens at saturating substrate concentration). The value of the Michaelis constant K_{M} is numerically equal to the substrate concentration at which the reaction rate is half of $V_{\max}$. Biochemical reactions involving a single substrate are often assumed to follow Michaelis–Menten kinetics, without regard to the model's underlying assumptions.

Model

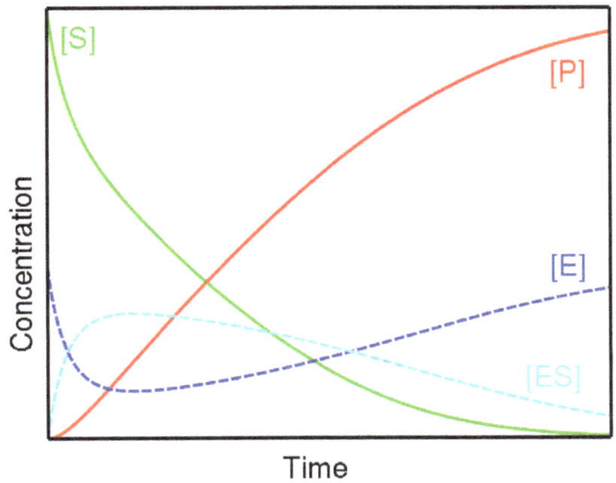

Change in concentrations over time for enzyme E, substrate S, complex ES and product P.

In 1901, French physical chemist Victor Henri found that enzyme reactions were initiated by a bond (more generally, a binding interaction) between the enzyme and the substrate. His work was taken up by German biochemist Leonor Michaelis and Canadian physician Maud Menten, who investigated the kinetics of an enzymatic reaction mechanism, invertase, that catalyzes the hydrolysis of sucrose into glucose and fructose. In 1913, they proposed a mathematical model of the reaction. It involves an enzyme, E, binding to a substrate, S, to form a complex, ES, which in turn releases a product, P, regenerating the original enzyme. This may be represented schematically as,

$$\mathrm{E} + \mathrm{S} \underset{k_r}{\overset{k_f}{\rightleftharpoons}} \mathrm{ES} \overset{k_{cat}}{\rightarrow} \mathrm{E} + \mathrm{P}$$

where k_f (forward rate constant), k_r (reverse rate constant), and k_{cat} (catalytic rate constant) denote the rate constants, the double arrows between S (substrate) and ES (enzyme-substrate complex) represent the fact that enzyme-substrate binding is a reversible process, and the single forward arrow represents the formation of P (product).

Under certain assumptions – such as the enzyme concentration being much less than the substrate concentration – the rate of product formation is given by,

$$v = \frac{d[P]}{dt} = V_{\max} \frac{[S]}{K_{\mathrm{M}} + [S]} = k_{\mathrm{cat}} [E]_0 \frac{[S]}{K_{\mathrm{M}} + [S]}.$$

The reaction order depends on the relative size of the two terms in the denominator. At low substrate concentration $[\mathrm{S}] \ll K_M$ so that $v = k_{\mathrm{cat}}[\mathrm{E}]_0 \frac{[\mathrm{S}]}{K_{\mathrm{M}}}$. Under these conditions the reaction rate varies linearly with substrate concentration [S] (first-order kinetics). However at higher [S] with $[S] \ll K_M$, the reaction becomes independent of [S] (zero-order kinetics) and asymptotically approaches its maximum rate $V_{max} = k_{cat}[E]_0$, where $[E]_0$ is the initial enzyme concentration. This rate is attained when all enzyme is bound to substrate. k_{cat}, the turnover number, is the maximum number of substrate molecules converted to product per enzyme molecule per second. Further addition of substrate does not increase the rate which is said to be saturated.

The value of the Michaelis constant K_M is numerically equal to the [S] at which the reaction rate is at half-maximum, and is an inverse measure of the substrate's affinity for the enzyme—as a small K_M indicates high affinity, meaning that the rate will approach V_{max} with lower [S] than those reactions with a larger K_M. The constant is not affected by the concentration or purity of an enzyme. The value of K_M is dependent on both the identity of enzyme and that of the substrate, as well as conditions such as temperature and pH.

The model is used in a variety of biochemical situations other than enzyme-substrate interaction, including antigen-antibody binding, DNA-DNA hybridization, and protein-protein interaction. It can be used to characterise a generic biochemical reaction, in the same way that the Langmuir equation can be used to model generic adsorption of biomolecular species. When an empirical equation of this form is applied to microbial growth, it is sometimes called a Monod equation.

Applications

Parameter values vary widely between enzymes:

Enzyme	K_M(M)	k_{cat}(s^{-1})	k_{cat}/K_M($M^{-1}s^{-1}$)
Chymotrypsin	1.5×10^{-2}	0.14	9.3
Pepsin	3.0×10^{-4}	0.50	1.7×10^{3}
T-RNA synthetase	9.0×10^{-4}	7.6	8.4×10^{3}
Ribonuclease	7.9×10^{-3}	7.9×10^{2}	1.0×10^{5}
Carbonic anhydrase	2.6×10^{-2}	4.0×10^{5}	1.5×10^{7}
Fumarase	5.0×10^{-6}	8.0×10^{2}	1.6×10^{8}

The constant k_{cat} / K_M (catalytic efficiency) is a measure of how efficiently an enzyme converts a substrate into product. Diffusion limited enzymes, such as fumarase, work at the theoretical upper limit of $10^8 - 10^{10}$ $M^{-1}s^{-1}$, limited by diffusion of substrate into the active site.

Michaelis–Menten kinetics have also been applied to a variety of spheres outside of biochemical reactions, including alveolar clearance of dusts, the richness of species pools, clearance of blood alcohol, the photosynthesis-irradiance relationship, and bacterial phage infection.

The equation can also be used to describe the relationship between ion channel conductivity and ligand concentration.

Derivation

Applying the law of mass action, which states that the rate of a reaction is proportional to the product of the concentrations of the reactants (i.e.[E][S]), gives a system of four non-linear ordinary differential equations that define the rate of change of reactants with time t,

$$\frac{d[\mathrm{E}]}{dt} = -k_f[\mathrm{E}][\mathrm{S}] + k_r[\mathrm{ES}] + k_{\mathrm{cat}}[\mathrm{ES}]$$

$$\frac{d[\mathrm{S}]}{dt} = -k_f[\mathrm{E}][\mathrm{S}] + k_r[\mathrm{ES}]$$

$$\frac{d[\mathrm{ES}]}{dt} = k_f[\mathrm{E}][\mathrm{S}] - k_r[\mathrm{ES}] - k_{\mathrm{cat}}[\mathrm{ES}]$$

$$\frac{d[\mathrm{P}]}{dt} = k_{\mathrm{cat}}[\mathrm{ES}].$$

In this mechanism, the enzyme E is a catalyst, which only facilitates the reaction, so that its total concentration, free plus combined, $[E]+[ES]=[E]_0$ is a constant. This conservation law can also be observed by adding the first and third equations above.

Equilibrium Approximation

In their original analysis, Michaelis and Menten assumed that the substrate is in instantaneous chemical equilibrium with the complex, which implies,

$$k_f[\mathrm{E}][\mathrm{S}] = k_r[\mathrm{ES}].$$

From the enzyme conservation law, we obtain,

$$[\mathrm{E}] = [\mathrm{E}]_0 - [\mathrm{ES}].$$

Combining the two expressions above, gives us,

$$k_f([\mathrm{E}]_0 - [\mathrm{ES}])[\mathrm{S}] = k_r[\mathrm{ES}].$$

Upon simplification we get,

$$[\mathrm{ES}] = \frac{[\mathrm{E}]_0[S]}{K_d + [\mathrm{S}]}$$

where $K_d = k_r / k_f$ is the dissociation constant for the enzyme-substrate complex. Hence the velocity v of the reaction – the rate at which P is formed – is,

$$v = \frac{d[\mathrm{P}]}{dt} = \frac{V_{\max}[S]}{K_d + [S]}$$

where $V_{\max} = k_{\mathrm{cat}}[\mathrm{E}]_0$ is the maximum reaction velocity.

Quasi-steady-state Approximation

An alternative analysis of the system was undertaken by British botanist G. E. Briggs and British geneticist J. B. S. Haldane in 1925. They assumed that the concentration of the intermediate complex does not change on the time-scale of product formation – known as the quasi-steady-state assumption or pseudo-steady-state-hypothesis. Mathematically, this assumption means $k_f[\mathrm{E}][\mathrm{S}] = k_r[\mathrm{ES}] + k_{\mathrm{cat}}[\mathrm{ES}]$. Combining this relationship with the enzyme conservation law, $[E] = [E]_0 - [ES]$, gives us,

$$\begin{aligned}(k_r + k_{\mathrm{cat}})[\mathrm{ES}] &= k_f[\mathrm{S}]([\mathrm{E}]_0 - [\mathrm{ES}]) \\ &= k_f[\mathrm{S}][\mathrm{E}]_0 - k_f[\mathrm{S}][\mathrm{ES}].\end{aligned}$$

Mozing $[\mathrm{ES}]$ to same side then factoring it out, the concentration of the complex is,

$$\begin{aligned}&= \frac{k_f[\mathrm{E}]_0[\mathrm{S}]}{(k_r + k_{\mathrm{cat}}) + k_f[\mathrm{S}]} \\ &= \frac{[\mathrm{E}]_0[\mathrm{S}]}{\left(\frac{k_r + k_{\mathrm{cat}}}{k_f}\right) + [\mathrm{S}]} \\ &= \frac{[\mathrm{E}]_0[\mathrm{S}]}{K_{\mathrm{M}} + [\mathrm{S}]}\end{aligned}$$

where,

$$K_{\mathrm{M}} = \frac{k_r + k_{\mathrm{cat}}}{k_f}$$

is known as the Michaelis constant, where k_r, k_{cat}, and k_f are, respectively, the constants for substrate unbinding, conversion to product, and binding to the enzyme. Hence the velocity v of the reaction is

$$v = \frac{d[\mathrm{P}]}{dt} = k_{\mathrm{cat}}[\mathrm{ES}] = \frac{V_{\max}[\mathrm{S}]}{K_{\mathrm{M}} + [\mathrm{S}]}$$

where,

$$V_{\max} = k_{\mathrm{cat}}[\mathrm{E}]_0.$$

Assumptions and Limitations

The first step in the derivation applies the law of mass action, which is reliant on free diffusion. However, in the environment of a living cell where there is a high concentration of proteins, the cytoplasm often behaves more like a gel than a liquid, limiting molecular movements and altering reaction rates. Although the law of mass action can be valid in heterogeneous environments, it is more appropriate to model the cytoplasm as a fractal, in order to capture its limited-mobility kinetics.

The resulting reaction rates predicted by the two approaches are similar, with the only difference being that the equilibrium approximation defines the constant as K_d, whilst the quasi-steady-state approximation uses K_M. However, each approach is founded upon a different assumption. The Michaelis–Menten equilibrium analysis is valid if the substrate reaches equilibrium on a much faster time-scale than the product is formed or, more precisely that,

$$\varepsilon_d = \frac{k_{\text{cat}}}{k_r} \ll 1.$$

By contrast, the Briggs–Haldane quasi-steady-state analysis is valid if,

$$\varepsilon_m = \frac{[\mathrm{E}]_0}{[\mathrm{S}]_0 + K_M} \ll 1.$$

Thus it holds if the enzyme concentration is much less than the substrate concentration or K_M or both.

In both the Michaelis–Menten and Briggs–Haldane analyses, the quality of the approximation improves as ε decreases. However, in model building, Michaelis–Menten kinetics are often invoked without regard to the underlying assumptions.

It is also important to remember that, while irreversibility is a necessary simplification in order to yield a tractable analytic solution, in the general case product formation is not in fact irreversible. The enzyme reaction is more correctly described as,

$$\mathrm{E} + \mathrm{S} \underset{k_{r1}}{\overset{k_{f1}}{\rightleftharpoons}} \mathrm{ES} \underset{k_{r2}}{\overset{k_{f2}}{\rightleftharpoons}} \mathrm{E} + \mathrm{P}\cdot$$

In general, the assumption of irreversibility is a good one in situations where one of the below is true:

1. The concentration of substrate(s) is very much larger than the concentration of products:

   $[\mathrm{S}] \gg [\mathrm{P}].$

This is true under standard *in vitro* assay conditions, and is true for many *in vivo* biological reactions, particularly where the product is continually removed by a subsequent reaction.

2. The energy released in the reaction is very large, that is,

   $\Delta G \ll 0.$

In situations where neither of these two conditions hold (that is, the reaction is low energy and a substantial pool of product(s) exists), the Michaelis–Menten equation breaks down, and more complex modelling approaches explicitly taking the forward and reverse reactions into account must be taken to understand the enzyme biology.

Determination of Constants

The typical method for determining the constants $V_{\max}$ and $V_{\max}$ involves running a series of enzyme assays at varying substrate concentrations [S], and measuring the initial reaction rate . 'Initial' here is taken to mean that the reaction rate is measured after a relatively short time period, during which it is assumed that the enzyme-substrate complex has formed, but that the substrate concentration held approximately constant, and so the equilibrium or quasi-steady-state approximation remain valid. By plotting reaction rate against concentration, and using nonlinear regression of the Michaelis–Menten equation, the parameters may be obtained.

Before computing facilities to perform nonlinear regression became available, graphical methods involving linearisation of the equation were used. A number of these were proposed, including the Eadie–Hofstee diagram, Hanes–Woolf plot and Lineweaver–Burk plot; of these, the Hanes–Woolf plot is the most accurate. However, while useful for visualization, all three methods distort the error structure of the data and are inferior to nonlinear regression. Assuming a similar error dv_0 on dv_0, an inverse representation leads to an error of dv_0 / v_0^2 on $1 / v_0$ (Propagation of uncertainty). Without proper estimation of dv_0 values, linearisation should be avoided. In addition, regression analysis using Least squares assumes that errors are normally distributed, which is not valid after a transformation of v_0 values. Nonetheless, their use can still be found in modern literature.

In 1997 Santiago Schnell and Claudio Mendoza suggested a closed form solution for the time course kinetics analysis of the Michaelis–Menten kinetics based on the solution of the Lambert W function. Namely,

$$\frac{[S]}{K_M} = W(F(t))$$

where W is the Lambert W function and

$$F(t) = \frac{[S]_0}{K_M} \exp\left(\frac{[S]_0}{K_M} - \frac{V_{\max}}{K_M} t\right).$$

The above equation has been used to estimate V_{max} and K_M from time course data.

Role of Substrate Unbinding

The Michaelis-Menten equation has been used to predict the rate of product formation in enzymatic reactions for more than a century. Specifically, it states that the rate of an enzymatic reaction will increase as substrate concentration increases, and that increased unbinding of enzyme-substrate complexes will decrease the reaction rate. While the first prediction is well established, the second is more elusive. Mathematical analysis of the effect of enzyme-substrate unbinding on enzymatic reactions at the single-molecule level has shown that unbinding of an enzyme from a substrate can reduce the rate of product formation under some conditions, but may also have the opposite effect. As substrate concentrations increase, a tipping point can be reached where an increase in the unbinding rate results in an increase, rather than a decrease, of the reaction rate. The results indicate that enzymatic reactions can behave in ways that violate the classical Michaelis-Menten equation, and that the role of unbinding in enzymatic catalysis still remains to be determined experimentally.

NON-MICHAELIS MENTEN KINETICS

Some enzymes display non-Michaelis-Menten kinetics that do not approximate to Michaelis-Menten kinetics to any useful extent. In such case there is little value in retaining the terminology and symbolism of Michaelis-Menten kinetics. Instead it is often possible to express the rate as a rational function of the substrate concentration:

$$v = \frac{\alpha_1[A] + \alpha_2[A]^2 + \alpha_3[A]^3 + \cdots + \alpha_n[A]^n}{1 + \beta_1[A] + \beta_2[A]^2 + \beta_3[A]^3 + \cdots + \beta_m[A]^m}$$

A rational function is a ratio of two polynomials. The degree of a polynomial is the largest exponent: thus, the degree of the numerator of the expression in Eqn is n and that of the denominator is m. The rational function as a whole may be described as a 'n:m function'. In general the degree of the numerator of a rate equation does not exceed the degree of the denominator for enzyme-catalysed reactions, but there is no other necessary relationship between n and m and neither bears any necessary relationship to the number of catalytic centres per molecule of enzyme. In the terminology of Eqn any rate equation obeying Michaelis-Menten kinetics can be defined as a 1:1 function.

Any coefficient α_i in the numerator of the right-hand side of Eqn has units $(\text{mol dm}^{-3})^{1-i}$ s^{-1}, and any coefficient β_i in the denominator has units $(\text{mol}^{-1}\text{ dm}^3)^i$. Similar equations are sometimes written in which the constant 1 in the denominator is replaced with a constant β_o. This practice is discouraged, because the equation then contains a redundant parameter and all of the coefficients α_i, β_i have undefined dimensions.

Under some conditions, which cannot be expressed simply and are not normally obvious from inspection of the coefficients of Eqn, the equation may generate a plot of v against [A] that shows a monotonic increase in v towards a limiting value at all positive finite values of [A]. A necessary, but not sufficient, condition is that the degrees of the numerator and denominator be equal, i.e. n = m. It is then meaningful to define a limiting rate $V = \alpha_n/\beta_n$. Moreover, it may also be useful to describe the kinetics quantitatively in terms of the slope of a plot of log[v /(V-v)] against log[A], which is known as a Hill plot. This slope is called the Hill coefficient and is given the symbol *h*. In a kinetic context it bears no necessary relationship of any kind to the number of catalytic centres per molecule of enzyme and it should not be given a symbol, such as n that suggests that it does. The symbol n H has sometimes been used as an alternative to h, but it should be borne in mind that this may cause difficulties in printing when used as an exponent. At any concentration of substrate at which h is greater than unity, the kinetics are said to display co-operativity. In some contexts the more explicit term positive co-operativity may be preferable to avoid ambiguity. At any concentration at which h = 1 the kinetics are said to be non-co-operative and if h is less than unity they are said to be negatively co-operative. In the case of Michaelis-Menten kinetics h = I over the whole concentration range, but in other cases h is not constant and the sign of co-operativity may change one or more times over the range of concentrations. Thus co-operativity is not absolute and in general can only be defined in relation to a particular concentration.

The term co-operativity is sometimes restricted to a purely mechanistic meaning, i.e. it is considered to refer to interactions between distinct sites on the enzyme. In common practice, however, the terms discussed above are frequently applied to enzymes in the absence of clear evidence for such interactions, and the aim therefore has been to legitimize such usage by providing purely operational definitions. In contexts where it is considered necessary to emphasize the operational character of the kind of co-operativity defined in terms of the Hill coefficient it may be qualified as Hill co-operativity or kinetic co-operativity.

References

- Anslyn, E.V.; Dougherty, D.A. (2006). Modern Physical Organic Chemistry. University Science Books. ISBN 978-1-891389-31-3
- Churchwell, M; Twaddle, N; Meeker, L; Doerge, D. (October 2005). “Improving Sensitivity in Liquid Chromatography-Mass Spectrometry”. Journal of Chromatography B. 825 (2): 134–143. doi:10.1016/j.jchromb.2005.05.037
- Enzyme-Kinetics, life-sciences: news-medical.net, Retrieved 24 April, 2019
- Ball, P. (2004). “Enzymes: By chance, or by design?”. Nature. 431 (7007): 396–397. Bibcode:-2004Natur.431..396B. doi:10.1038/431396a. PMID 15385982

Enzyme Inhibition

4 CHAPTER

The process through which the catalytic efficiency of an enzyme is decreased is known as enzyme inhibition. The molecules which cause inhibition in enzymes are known as inhibitors. A few types of inhibitors are uncompetitive inhibitors and targeted covalent inhibitors. All the diverse processes related to enzyme inhibition have been carefully analyzed in this chapter.

Enzymes can be regulated in ways that either promote or reduce their activity. There are many different kinds of molecules that inhibit or promote enzyme function, and various mechanisms exist for doing so. In some cases of enzyme inhibition, for example, an inhibitor molecule is similar enough to a substrate that it can bind to the active site and simply block the substrate from binding. When this happens, the enzyme is inhibited through competitive inhibition, because an inhibitor molecule competes with the substrate for active site binding. On the other hand, in noncompetitive inhibition, an inhibitor molecule binds to the enzyme in a location other than an allosteric site and still manages to block substrate binding to the active site.

Elucidating Mechanisms for the Inhibition of Enzyme Catalysis

When an inhibitor interacts with an enzyme it decreases the enzyme's catalytic efficiency. An irreversible inhibitor covalently binds to the enzyme's active site, producing a permanent loss in catalytic efficiency even if we decrease the inhibitor's concentration. A reversible inhibitor forms a noncovalent complex with the enzyme, resulting in a temporary decrease in catalytic efficiency. If we remove the inhibitor, the enzyme's catalytic efficiency returns to its normal level.

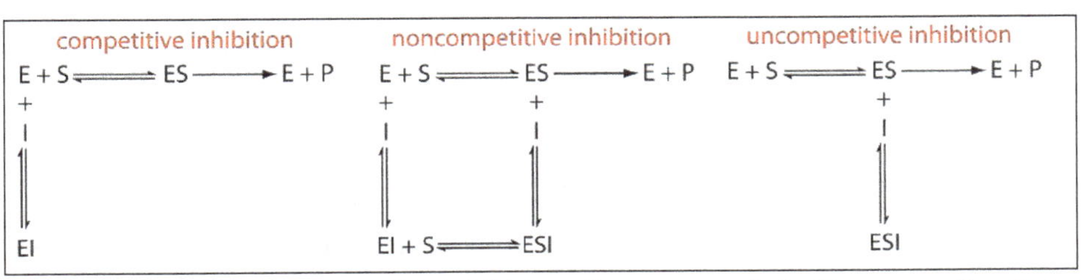

Figure: Mechanisms for the reversible inhibition of enzyme catalysis. E: enzyme, S: substrate, P: product, I: inhibitor, ES: enzyme–substrate complex, EI: enzyme–inhibitor complex, ESI: enzyme–substrate–inhibitor complex.

There are several pathways for the reversible binding of an inhibitor to an enzyme, in figure. In competitive inhibition the substrate and the inhibitor compete for the same active site on the enzyme. Because the substrate cannot bind to an enzyme–inhibitor complex, EI, the enzyme's catalytic efficiency for the substrate decreases. With noncompetitive inhibition the substrate and the inhibitor bind to different active sites on the enzyme, forming an enzyme–substrate–inhibitor, or ESI complex. The formation of an ESI complex decreases catalytic efficiency because only the enzyme–substrate complex reacts to form the product. Finally, in uncompetitive inhibition the inhibitor binds to the enzyme–substrate complex, forming an inactive ESI complex.

We can identify the type of reversible inhibition by observing how a change in the inhibitor's concentration affects the relationship between the rate of reaction and the substrate's concentration. When we display kinetic data using as a Lineweaver-Burk plot it is easy to determine which mechanism is in effect. For example, an increase in slope, a decrease in the x-intercept, and no change in the y-intercept indicates competitive inhibition. Because the inhibitor's binding is reversible, we can still obtain the same maximum velocity—thus the constant value for the y-intercept—by adding enough substrate to completely displace the inhibitor. Because it takes more substrate, the value of Km increases, which explains the increase in the slope and the decrease in the x-intercept's value.

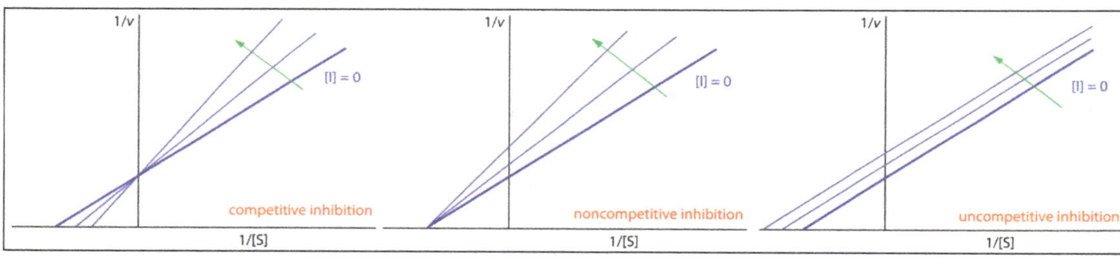

Figure: Linweaver–Burk plots for competitive inhibition, noncompetitive inhibition, and uncompetitive inhibition. The thick blue line in each plot shows the kinetic behavior in the absence of inhibitor, and the thin blue lines in each plot show the change in behavior for increasing concentrations of the inhibitor. In each plot, the inhibitor's concentration increases in the direction of the green arrow.

ENZYME INHIBITOR

An enzyme inhibitor is a molecule that binds to an enzyme and decreases its activity. Since blocking an enzyme's activity can kill a pathogen or correct a metabolic imbalance, many drugs are enzyme inhibitors. They are also used in pesticides. Not all molecules that bind to enzymes are inhibitors; *enzyme activators* bind to enzymes and increase their enzymatic activity, while enzyme substrates bind and are converted to products in the normal catalytic cycle of the enzyme.

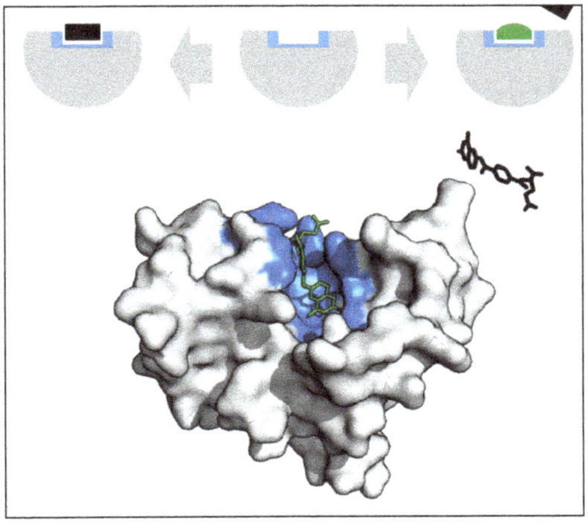

An enzyme binding site that would normally bind substrate can alternatively bind a competitive inhibitor, preventing substrate access. Dihydrofolate reductase is inhibited by methotrexate which prevents binding of its substrate, folic acid. Binding site in blue, inhibitor in green, and substrate in black.

The binding of an inhibitor can stop a substrate from entering the enzyme's active site and/or hinder the enzyme from catalyzing its reaction. Inhibitor binding is either reversible or irreversible. Irreversible inhibitors usually react with the enzyme and change it chemically (e.g. via covalent bond formation). These inhibitors modify key amino acid residues needed for enzymatic activity. In contrast, reversible inhibitors bind non-covalently and different types of inhibition are produced depending on whether these inhibitors bind to the enzyme, the enzyme-substrate complex, or both.

Many drug molecules are enzyme inhibitors, so their discovery and improvement is an active area of research in biochemistry and pharmacology. A medicinal enzyme inhibitor is often judged by its specificity (its lack of binding to other proteins) and its potency (its dissociation constant, which indicates the concentration needed to inhibit the enzyme). A high specificity and potency ensure that a drug will have few side effects and thus low toxicity.

Enzyme inhibitors also occur naturally and are involved in the regulation of metabolism. For example, enzymes in a metabolic pathway can be inhibited by downstream products. This type of negative feedback slows the production line when products begin to build up and is an important way to maintain homeostasis in a cell. Other cellular enzyme inhibitors are proteins that specifically bind to and inhibit an enzyme target. This can help control enzymes that may be damaging to a cell, like proteases or nucleases. A well-characterised example of this is the ribonuclease inhibitor, which binds to ribonucleases in one of the tightest known protein–protein interactions. Natural enzyme inhibitors can also be poisons and are used as defenses against predators or as ways of killing prey.

Reversible Inhibitors

Types of Reversible Inhibitors

Reversible inhibitors attach to enzymes with non-covalent interactions such as hydrogen bonds, hydrophobic interactions and ionic bonds. Multiple weak bonds between the inhibitor and the active site combine to produce strong and specific binding. In contrast to substrates and irreversible inhibitors, reversible inhibitors generally do not undergo chemical reactions when bound to the enzyme and can be easily removed by dilution or dialysis.

There are four kinds of reversible enzyme inhibitors. They are classified according to the effect of varying the concentration of the enzyme's substrate on the inhibitor.

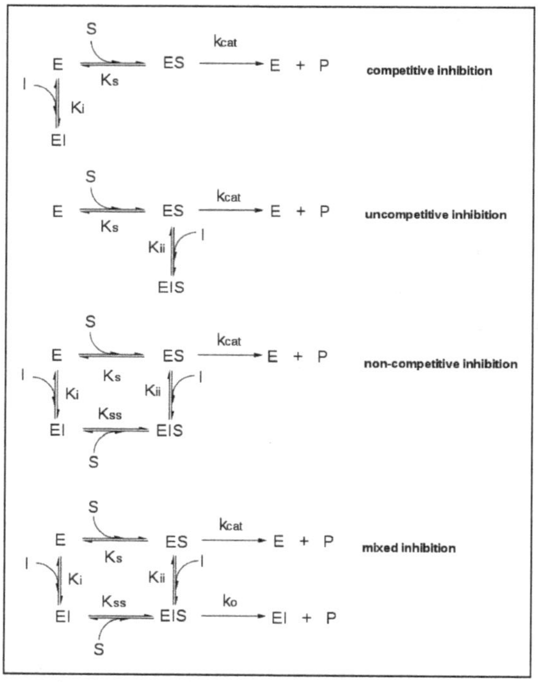

Types of inhibition.

- In competitive inhibition, the substrate and inhibitor cannot bind to the enzyme at the same time. This usually results from the inhibitor having an affinity for the active site of an enzyme where the substrate also binds; the substrate and inhibitor *compete* for access to the enzyme's active site. This type of inhibition can be overcome by sufficiently high concentrations of substrate (V_{max} remains constant), i.e., by out-competing the inhibitor. However, the apparent K_m will increase as it takes a higher concentration of the substrate to reach the K_m point, or half the V_{max}. Competitive inhibitors are often similar in structure to the real substrate.

- In uncompetitive inhibition, the inhibitor binds only to the substrate-enzyme complex. This type of inhibition causes V_{max} to decrease (maximum velocity decreases as a result of removing activated complex) and K_m to decrease (due to better binding efficiency as a result of Le Chatelier's principle and the effective elimination of the ES complex thus decreasing the K_m which indicates a higher binding affinity).

- In non-competitive inhibition, the binding of the inhibitor to the enzyme reduces its activity but does not affect the binding of substrate. As a result, the extent of inhibition depends only on the concentration of the inhibitor. V_{max} will decrease due to the inability for the reaction to proceed as efficiently, but K_m will remain the same as the actual binding of the substrate, by definition, will still function properly.
- In mixed inhibition, the inhibitor can bind to the enzyme at the same time as the enzyme's substrate. However, the binding of the inhibitor affects the binding of the substrate, and vice versa. This type of inhibition can be reduced, but not overcome by increasing concentrations of substrate. Although it is possible for mixed-type inhibitors to bind in the active site, this type of inhibition generally results from an allosteric effect where the inhibitor binds to a different site on an enzyme. Inhibitor binding to this allosteric site changes the conformation (i.e., tertiary structure or three-dimensional shape) of the enzyme so that the affinity of the substrate for the active site is reduced.

Quantitative Description of Reversible Inhibition

Reversible inhibition can be described quantitatively in terms of the inhibitor's binding to the enzyme and to the enzyme-substrate complex, and its effects on the kinetic constants of the enzyme. In the classic Michaelis-Menten scheme, an enzyme (E) binds to its substrate (S) to form the enzyme–substrate complex ES. Upon catalysis, this complex breaks down to release product P and free enzyme. The inhibitor (I) can bind to either E or ES with the dissociation constants K_i or K_i', respectively.

- Competitive inhibitors can bind to E, but not to ES. Competitive inhibition increases K_m (i.e., the inhibitor interferes with substrate binding), but does not affect V_{max} (the inhibitor does not hamper catalysis in ES because it cannot bind to ES).
- Uncompetitive inhibitors bind to ES. Uncompetitive inhibition decreases both K_m' and 'V_{max}. *The inhibitor affects substrate binding by increasing the enzyme's affinity for the substrate (decreasing* K_m*) as well as hampering catalysis (decreases* V_{max}*).*
- Non-competitive inhibitors have identical affinities for E and ES ($K_i = K_i'$). Non-competitive inhibition does not change K_m (i.e., it does not affect substrate binding) but decreases V_{max} (i.e., inhibitor binding hampers catalysis).

- Mixed-type inhibitors bind to both E and ES, but their affinities for these two forms of the enzyme are different ($K_i \neq K_i$'). Thus, mixed-type inhibitors interfere with substrate binding (increase K_m) and hamper catalysis in the ES complex (decrease V_{max}).

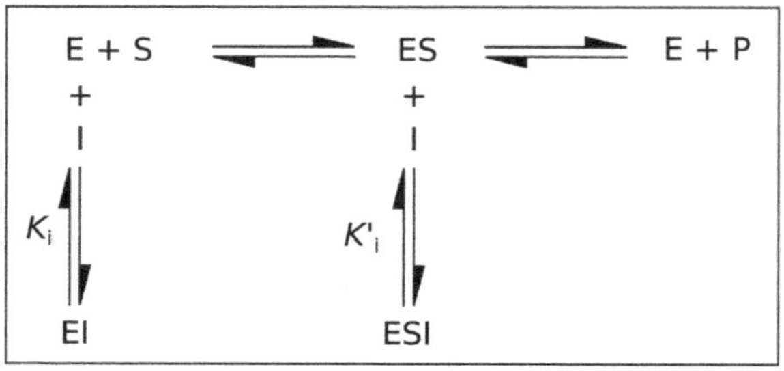

Kinetic scheme for reversible enzyme inhibitors.

When an enzyme has multiple substrates, inhibitors can show different types of inhibition depending on which substrate is considered. This results from the active site containing two different binding sites within the active site, one for each substrate. For example, an inhibitor might compete with substrate A for the first binding site, but be a non-competitive inhibitor with respect to substrate B in the second binding site.

Measuring the Dissociation Constants of a Reversible Inhibitor

An enzyme inhibitor is characterised by its two dissociation constants, K_i and K_i', to the enzyme and to the enzyme-substrate complex, respectively. The enzyme-inhibitor constant K_i can be measured directly by various methods; one extremely accurate method is isothermal titration calorimetry, in which the inhibitor is titrated into a solution of enzyme and the heat released or absorbed is measured. However, the other dissociation constant K_i' is difficult to measure directly, since the enzyme-substrate complex is short-lived and undergoing a chemical reaction to form the product. Hence, K_i' is usually measured indirectly, by observing the enzyme activity under various substrate and inhibitor concentrations, and fitting the data to a modified Michaelis–Menten equation:

$$V = \frac{V_{max}[S]}{\alpha K_m + \alpha'[S]} = \frac{(1/\alpha')V_{max}[S]}{(\alpha/\alpha')K_m + [S]}$$

where the modifying factors α and α' are defined by the inhibitor concentration and its two dissociation constants,

$$\alpha = 1 + \frac{[I]}{K_i}$$

$$\alpha' = 1 + \frac{[I]}{K_i'}.$$

Thus, in the presence of the inhibitor, the enzyme's effective K_m and V_{max} become (α/α') K_m and $(1/\alpha')V_{max}$, respectively. However, the modified Michaelis-Menten equation assumes that binding of the inhibitor to the enzyme has reached equilibrium, which may be a very slow process for inhibitors with sub-nanomolar dissociation constants. In these cases, it is usually more practical to treat the tight-binding inhibitor as an irreversible inhibitor; however, it can still be possible to estimate K_i' kinetically if K_i is measured independently.

The effects of different types of reversible enzyme inhibitors on enzymatic activity can be visualized using graphical representations of the Michaelis–Menten equation, such as Lineweaver–Burk and Eadie-Hofstee plots. For example, in the Lineweaver–Burk plots at the right, the competitive inhibition lines intersect on the y-axis, illustrating that such inhibitors do not affect V_{max}. Similarly, the non-competitive inhibition lines intersect on the x-axis, showing these inhibitors do not affect K_m. However, it can be difficult to estimate K_i and K_i' accurately from such plots, so it is advisable to estimate these constants using more reliable nonlinear regression methods.

Reversible Inhibitors

Traditionally reversible enzyme inhibitors have been classified as competitive, uncompetitive, or non-competitive, according to their effects on K_m and V_{max}. These different effects result from the inhibitor binding to the enzyme E, to the enzyme–substrate complex ES, or to both, respectively. The division of these classes arises from a problem in their derivation and results in the need to use two different binding constants for one binding event. The binding of an inhibitor and its effect on the enzymatic activity are two distinctly different things, another problem the traditional equations fail to acknowledge. In noncompetitive inhibition the binding of the inhibitor results in 100% inhibition of the enzyme only, and fails to consider the possibility of anything in between. The common form of the inhibitory term also obscures the relationship between the inhibitor binding to the enzyme and its relationship to any other binding term be it the Michaelis–Menten equation or a dose response curve associated with ligand receptor binding. To demonstrate the relationship the following rearrangement can be made:

$$\frac{V_{max}}{1+\frac{[\mathrm{I}]}{K_i}} = V_{max}\left(\frac{K_i}{K_i+[\mathrm{I}]}\right) \qquad \text{multiply by } \frac{K_i}{K_i}=1$$

$$= V_{max}\left(\frac{K_i+[\mathrm{I}]-[\mathrm{I}]}{K_i+[\mathrm{I}]}\right) \qquad \text{add } [\mathrm{I}]-[\mathrm{I}]=0 \text{ to numerator}$$

$$= V_{max}\left(1-\frac{[\mathrm{I}]}{K_i+[\mathrm{I}]}\right) \qquad \text{simplify } \frac{K_i+[\mathrm{I}]}{K_i+[\mathrm{I}]}=1$$

$$= V_{max} - V_{max}\frac{[\mathrm{I}]}{K_i+[\mathrm{I}]} \qquad \text{multiply out by } V_{max}$$

This rearrangement demonstrates that similar to the Michaelis–Menten equation, the maximal rate of reaction depends on the proportion of the enzyme population interacting with its substrate.

fraction of the enzyme population bound by substrate,

$$\frac{[\mathrm{S}]}{[\mathrm{S}]+K_m}$$

fraction of the enzyme population bound by inhibitor,

$$\frac{[\mathrm{I}]}{[\mathrm{I}]+K_i}$$

The effect of the inhibitor is a result of the percent of the enzyme population interacting with inhibitor. The only problem with this equation in its present form is that it assumes absolute inhibition of the enzyme with inhibitor binding, when in fact there can be a wide range of effects anywhere from 100% inhibition of substrate turn over to just >0%. To account for this the equation can be easily modified to allow for different degrees of inhibition by including a delta V_{max} term.

$$V_{max} - \Delta V_{max}\frac{[\mathrm{I}]}{[\mathrm{I}]+K_i}$$

or

$$V_{max1} - (V_{max1} - V_{max2})\frac{[\mathrm{I}]}{[\mathrm{I}]+K_i}$$

This term can then define the residual enzymatic activity present when the inhibitor is interacting with individual enzymes in the population. However the inclusion of this term has the added value of allowing for the possibility of activation if the secondary V_{max} term turns out to be higher than the initial term. To account for the possibly of activation as well the notation can then be rewritten replacing the inhibitor "I" with a modifier term denoted here as "X".

$$V_{max1} - (V_{max1} - V_{max2})\frac{[\mathrm{X}]}{[\mathrm{X}]+K_x}$$

While this terminology results in a simplified way of dealing with kinetic effects relating to the maximum velocity of the Michaelis–Menten equation, it highlights potential problems with the term used to describe effects relating to the K_m. The K_m relating to the affinity of the enzyme for the substrate should in most cases relate to potential changes in the binding site of the enzyme which would directly result from enzyme inhibitor interactions. As such a term similar to the one proposed above to modulate V_{max} should be appropriate in most situations:

$$K_{m1} - (K_{m1} - K_{m2})\frac{[X]}{[X]+K_x}$$

Special Cases

- The mechanism of partially competitive inhibition is similar to that of non-competitive, except that the EIS complex has catalytic activity, which may be lower or even higher (partially competitive activation) than that of the enzyme–substrate (ES) complex. This inhibition typically displays a lower V_{max}, but an unaffected K_m value.
- Uncompetitive inhibition occurs when the inhibitor binds only to the enzyme–substrate complex, not to the free enzyme; the EIS complex is catalytically inactive. This mode of inhibition is rare and causes a decrease in both V_{max} and the K_m value.
- Substrate and product inhibition is where either the substrate or product of an enzyme reaction inhibit the enzyme's activity. This inhibition may follow the competitive, uncompetitive or mixed patterns. In substrate inhibition there is a progressive decrease in activity at high substrate concentrations. This may indicate the existence of two substrate-binding sites in the enzyme. At low substrate, the high-affinity site is occupied and normal kinetics are followed. However, at higher concentrations, the second inhibitory site becomes occupied, inhibiting the enzyme. Product inhibition is often a regulatory feature in metabolism and can be a form of negative feedback.
- Slow-tight inhibition occurs when the initial enzyme–inhibitor complex EI undergoes isomerisation to a second more tightly held complex, EI*, but the overall inhibition process is reversible. This manifests itself as slowly increasing enzyme inhibition. Under these conditions, traditional Michaelis–Menten kinetics give a false value for K_i, which is time–dependent. The true value of K_i can be obtained through more complex analysis of the on (k_{on}) and off (k_{off}) rate constants for inhibitor association.

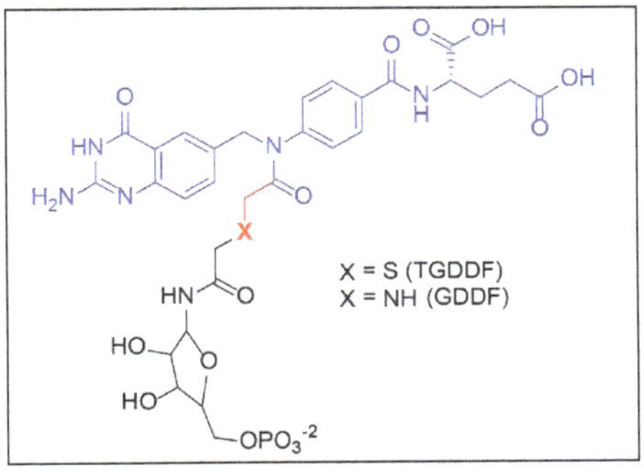

TGDDF / GDDF MAIs where blue depicts the tetrahydrofolate cofactor analog, black GAR or thioGAR and red, the connecting atoms.

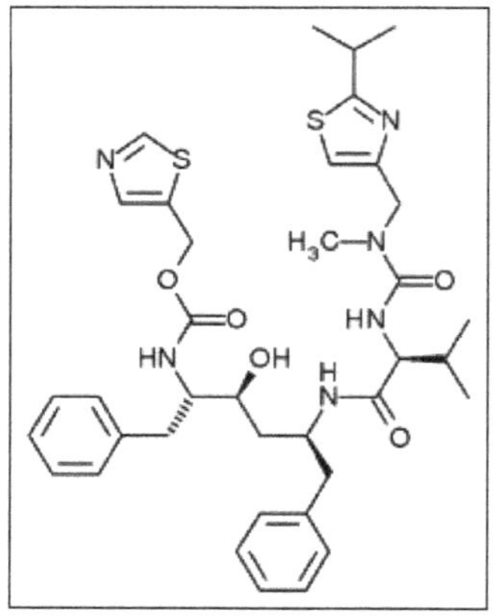

Peptide-based HIV-1 protease inhibitor ritonavir.

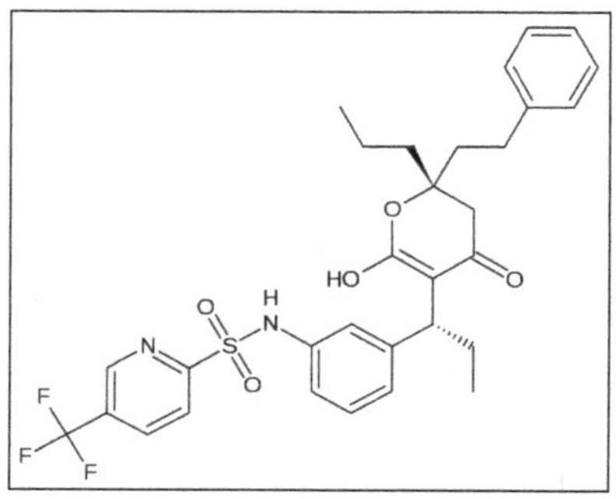

Nonpeptidic HIV-1 protease inhibitor tipranavir.

- Bi-substrate analog inhibitors are high affinity and selectivity inhibitors that can be prepared for enzymes that catalyze bi-molecular reactions by capturing the binding energy of each substrate into one molecule. For example, in the formyl transfer reactions of purine biosynthesis, a potent multi-substrate adduct inhibitor (MAI) to GAR TFase was prepared synthetically by linking analogs of the glycinamide ribonucleotide (GAR) substrate and the N-10-formyl tetrahydrofolate cofactor together to produce thioglycinamide ribonucleotide dideazafolate (TGDDF), or enzymatically from the natural GAR substrate to yield GDDF. Here the subnanomolar dissociation constant (KD) of TGDDF was greater than predicted presumably due to entropic advantages gained and/or positive interactions acquired through the atoms linking the components. MAIs have also been observed to be produced in cells by reactions of pro-drugs such as isoniazid or enzyme inhibitor ligands (e.g., PTC124) with cellular cofactors such as NADH and ATP respectively.

Examples of Reversible Inhibitors

As enzymes have evolved to bind their substrates tightly, and most reversible inhibitors bind in the active site of enzymes, it is unsurprising that some of these inhibitors are strikingly similar in structure to the substrates of their targets. Inhibitors of DHFR are prominent examples. Other example of these substrate mimics are the protease inhibitors, a very successful class of antiretroviral drugs used to treat HIV. The structure of ritonavir, a protease inhibitor based on a peptide and containing three peptide bonds, is shown on the right. As this drug resembles the protein that is the substrate of the HIV protease, it competes with this substrate in the enzyme's active site.

Enzyme inhibitors are often designed to mimic the transition state or intermediate of an enzyme-catalyzed reaction. This ensures that the inhibitor exploits the transition state stabilising effect of the enzyme, resulting in a better binding affinity (lower K_i) than substrate-based designs. An example of such a transition state inhibitor is the antiviral drug oseltamivir; this drug mimics the planar nature of the ring oxonium ion in the reaction of the viral enzyme neuraminidase.

However, not all inhibitors are based on the structures of substrates. For example, the structure of another HIV protease inhibitor tipranavir is shown on the left. This molecule is not based on a peptide and has no obvious structural similarity to a protein substrate. These non-peptide inhibitors can be more stable than inhibitors containing peptide bonds, because they will not be substrates for peptidases and are less likely to be degraded.

In drug design it is important to consider the concentrations of substrates to which the target enzymes are exposed. For example, some protein kinase inhibitors have chemical structures that are similar to adenosine triphosphate, one of the substrates of these enzymes. However, drugs that are simple competitive inhibitors will have to compete with the high concentrations of ATP in the cell. Protein kinases can also be inhibited by competition at the binding sites where the kinases interact with their substrate proteins, and most proteins are present inside cells at concentrations much lower than the concentration of ATP. As a consequence, if two protein kinase inhibitors both bind in the active site with similar affinity, but only one has to compete with ATP, then the competitive inhibitor at the protein-binding site will inhibit the enzyme more effectively.

Irreversible Inhibitors

Types of Irreversible Inhibition (Covalent Inactivation)

Irreversible inhibitors usually covalently modify an enzyme, and inhibition can therefore not be reversed. Irreversible inhibitors often contain reactive functional groups such as nitrogen mustards, aldehydes, haloalkanes, alkenes, Michael acceptors, phenyl sulfonates, or fluorophosphonates. These electrophilic groups react with amino acid side chains to form covalent adducts. The residues modified are those with side chains containing nucleophiles such as hydroxyl or sulfhydryl groups; these include the amino acids serine (as in DFP, right), cysteine, threonine, or tyrosine.

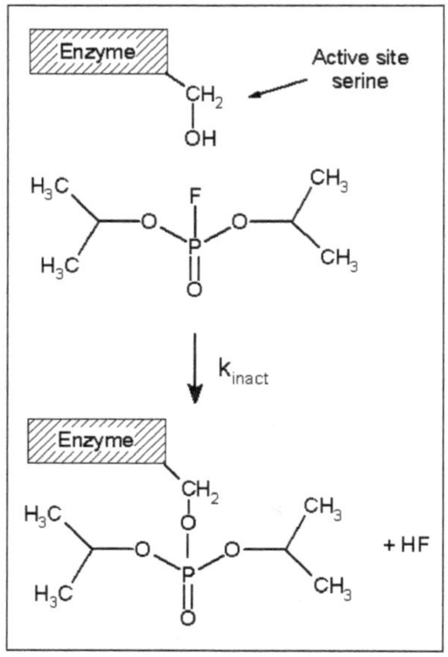

Reaction of the irreversible inhibitor diisopropylfluorophosphate (DFP) with a serine protease.

Irreversible inhibition is different from irreversible enzyme inactivation. Irreversible inhibitors are generally specific for one class of enzyme and do not inactivate all proteins; they do not function by destroying protein structure but by specifically altering the active site of their target. For example, extremes of pH or temperature usually cause denaturation of all protein structure, but this is a non-specific effect. Similarly, some non-specific chemical treatments destroy protein structure: for example, heating in concentrated hydrochloric acid will hydrolyse the peptide bonds holding proteins together, releasing free amino acids.

Irreversible inhibitors display time-dependent inhibition and their potency therefore cannot be characterised by an IC_{50} value. This is because the amount of active enzyme at a given concentration of irreversible inhibitor will be different depending on how long the inhibitor is pre-incubated with the enzyme. Instead, $k_{obs}/[I]$ values are used, where k_{obs} is the observed pseudo-first order rate of inactivation (obtained by plotting the log of % activity vs. time) and $[I]$ is the concentration of inhibitor. The $k_{obs}/[I]$ parameter is valid as long as the inhibitor does not saturate binding with the enzyme (in which case $k_{obs} = k_{inact}$).

Analysis of Irreversible Inhibition

Irreversible inhibitors have a short instance where they form a reversible non-covalent complex with the enzyme (EI or ESI) and this then reacts to produce the covalently modified "dead-end complex" EI* (an irreversible covalent complex). The rate at which EI* is formed is called the inactivation rate or k_{inact}. Since formation of EI may compete with ES, binding of irreversible inhibitors can be prevented by competition either with substrate or with a second, reversible inhibitor. This protection effect is good evidence of a specific reaction of the irreversible inhibitor with the active site.

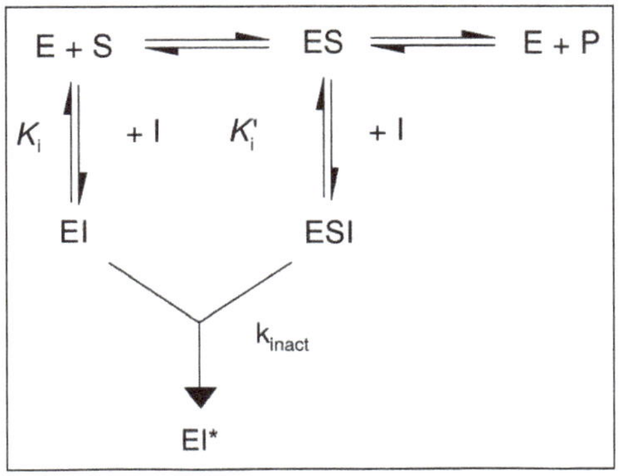

Kinetic scheme for irreversible inhibitors.

The binding and inactivation steps of this reaction are investigated by incubating the enzyme with inhibitor and assaying the amount of activity remaining over time. The activity will be decreased in a time-dependent manner, usually following exponential decay. Fitting these data to a rate equation gives the rate of inactivation at this concentration of inhibitor. This is done at several different concentrations of inhibitor. If a reversible EI complex is involved the inactivation rate will be saturable and fitting this curve will give k_{inact} and K_i.

Another method that is widely used in these analyses is mass spectrometry. Here, accurate measurement of the mass of the unmodified native enzyme and the inactivated enzyme gives the increase in mass caused by reaction with the inhibitor and shows the stoichiometry of the reaction. This is usually done using a MALDI-TOF mass spectrometer. In a complementary technique, peptide mass fingerprinting involves digestion of the native and modified protein with a protease such as trypsin. This will produce a set of peptides that can be analysed using a mass spectrometer. The peptide that changes in mass after reaction with the inhibitor will be the one that contains the site of modification.

Special Cases

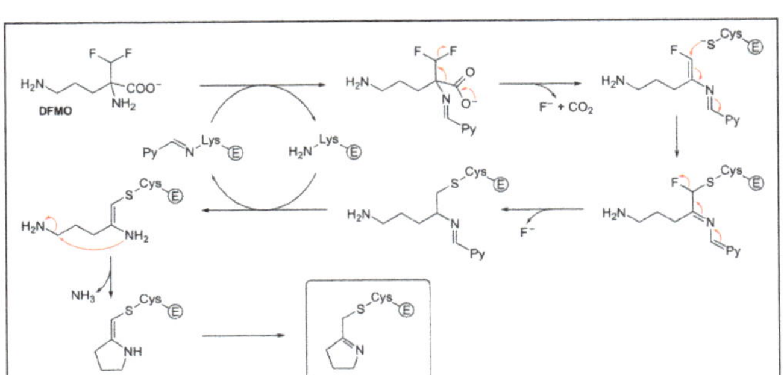

Chemical mechanism for irreversible inhibition of ornithine decarboxylase by DFMO. Pyridoxal 5'-phosphate (Py) and enzyme (E) are not shown.

Not all irreversible inhibitors form covalent adducts with their enzyme targets. Some reversible inhibitors bind so tightly to their target enzyme that they are essentially irreversible. These tight-binding inhibitors may show kinetics similar to covalent irreversible inhibitors. In these cases, some of these inhibitors rapidly bind to the enzyme in a low-affinity EI complex and this then undergoes a slower rearrangement to a very tightly bound EI* complex. This kinetic behaviour is called slow-binding. This slow rearrangement after binding often involves a conformational change as the enzyme "clamps down" around the inhibitor molecule. Examples of slow-binding inhibitors include some important drugs, such methotrexate, allopurinol, and the activated form of acyclovir.

Examples of Irreversible Inhibitors

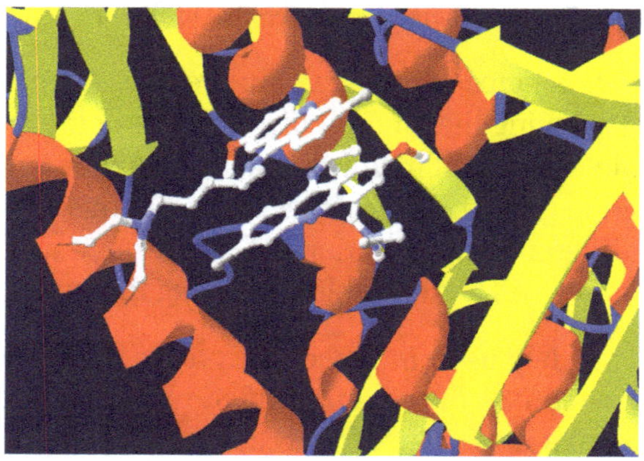

Trypanothione reductase with the lower molecule of an inhibitor bound irreversibly and the upper one reversibly.

Diisopropylfluorophosphate (DFP) is shown as an example of an irreversible protease inhibitor. The enzyme hydrolyses the phosphorus–fluorine bond, but the phosphate residue remains bound to the serine in the active site, deactivating it. Similarly, DFP also reacts with the active site of acetylcholine esterase in the synapses of neurons, and consequently is a potent neurotoxin, with a lethal dose of less than 100 mg.

Suicide inhibition is an unusual type of irreversible inhibition where the enzyme converts the inhibitor into a reactive form in its active site. An example is the inhibitor of polyamine biosynthesis, α-difluoromethylornithine or DFMO, which is an analogue of the amino acid ornithine, and is used to treat African trypanosomiasis (sleeping sickness). Ornithine decarboxylase can catalyse the decarboxylation of DFMO instead of ornithine. However, this decarboxylation reaction is followed by the elimination of a fluorine atom, which converts this catalytic intermediate into a conjugated imine, a highly electrophilic species. This reactive form of DFMO then reacts with either a cysteine or lysine residue in the active site to irreversibly inactivate the enzyme.

Since irreversible inhibition often involves the initial formation of a non-covalent EI complex, it is sometimes possible for an inhibitor to bind to an enzyme in more than one way. For example, trypanothione reductase from the human protozoan parasite *Trypanosoma cruzi*, two molecules of an inhibitor called *quinacrine mustard* are bound in its active site. The top molecule is bound reversibly, but the lower one is bound covalently as it has reacted with an amino acid residue through its nitrogen mustard group.

Discovery and Design of Inhibitors

Robots used for the high-throughput screening of chemical libraries to discover new enzyme inhibitors.

New drugs are the products of a long drug development process, the first step of which is often the discovery of a new enzyme inhibitor. In the past the only way to discover these new inhibitors was by trial and error: Screening huge libraries of compounds against a target enzyme and hoping that some useful leads would emerge. This brute force approach is still successful and has even been extended by combinatorial chemistry approaches that quickly produce large numbers of novel compounds and high-throughput screening technology to rapidly screen these huge chemical libraries for useful inhibitors.

More recently, an alternative approach has been applied: rational drug design uses the three-dimensional structure of an enzyme's active site to predict which molecules might be inhibitors. These predictions are then tested and one of these tested compounds may be a novel inhibitor. This new inhibitor is then used to try to obtain a structure of the enzyme in an inhibitor/enzyme complex to show how the molecule is binding to the active site, allowing changes to be made to the inhibitor to try to optimise binding. This test and improve cycle is then repeated until a sufficiently potent inhibitor is produced. Computer-based methods of predicting the affinity of an inhibitor for an enzyme are also being developed, such as molecular docking and molecular mechanics.

Uses of Inhibitors

Enzyme inhibitors are found in nature and are also designed and produced as part of pharmacology and biochemistry. Natural poisons are often enzyme inhibitors that have evolved to defend a plant or animal against predators. These natural toxins include some of the most poisonous compounds known. Artificial inhibitors are often used as drugs, but can also be insecticides such as malathion, herbicides such as glyphosate, or disinfectants such as triclosan. Other artificial enzyme inhibitors block acetylcholinesterase, an enzyme which breaks down acetylcholine, and are used as nerve agents in chemical warfare.

Chemotherapy

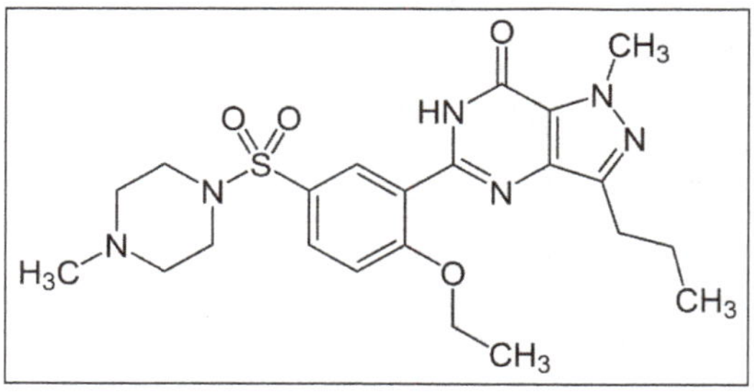

The structure of sildenafil (Viagra).

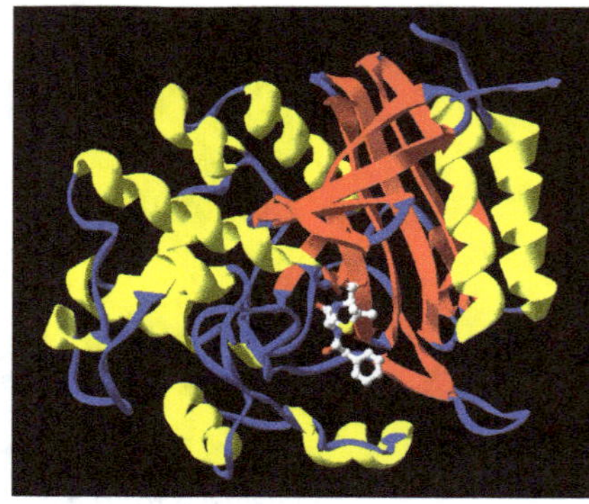

The structure of a complex between penicillin G and the *Streptomyces* transpeptidase.

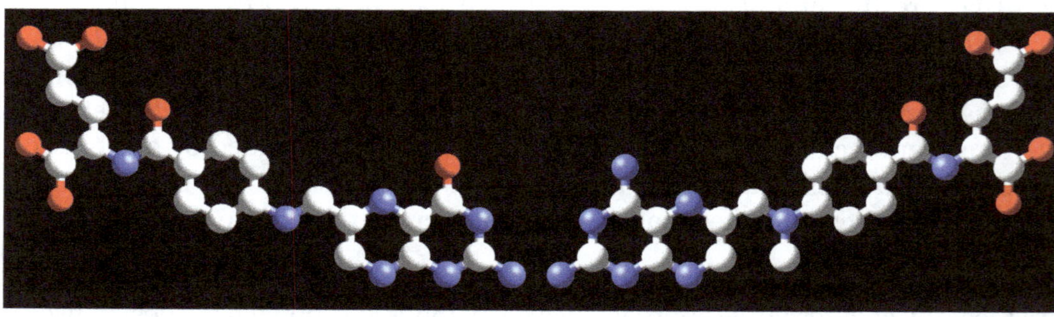

The coenzyme folic acid (left) compared to the anti-cancer drug methotrexate (right).

The most common uses for enzyme inhibitors are as drugs to treat disease. Many of these inhibitors target a human enzyme and aim to correct a pathological condition. However, not all drugs are enzyme inhibitors. Some, such as anti-epileptic drugs, alter enzyme activity by causing more or less of the enzyme to be produced. These effects are called enzyme induction and inhibition and are alterations in gene expression, which is unrelated to the type of enzyme inhibition. Other drugs interact with cellular targets that are not enzymes, such as ion channels or membrane receptors.

An example of a medicinal enzyme inhibitor is sildenafil (Viagra), a common treatment for male erectile dysfunction. This compound is a potent inhibitor of cGMP specific phosphodiesterase type 5, the enzyme that degrades the signalling molecule cyclic guanosine monophosphate. This signalling molecule triggers smooth muscle relaxation and allows blood flow into the corpus cavernosum, which causes an erection. Since the drug decreases the activity of the enzyme that halts the signal, it makes this signal last for a longer period of time.

Another example of the structural similarity of some inhibitors to the substrates of the enzymes they target is In the figure, a molecule of penicillincomparing the drug methotrexate to folic acid. Folic acid is a substrate of dihydrofolate reductase, an enzyme involved in making nucleotides that is potently inhibited by methotrexate. Methotrexate blocks the action of dihydrofolate reductase and thereby halts the production of nucleotides. This block of nucleotide biosynthesis is more toxic to rapidly growing cells than non-dividing cells, since a rapidly growing cell has to carry out DNA replication, therefore methotrexate is often used in cancer chemotherapy.

Antibiotics

Drugs also are used to inhibit enzymes needed for the survival of pathogens. For example, bacteria are surrounded by a thick cell wall made of a net-like polymer called peptidoglycan. Many antibiotics such as penicillin and vancomycin inhibit the enzymes that produce and then cross-link the strands of this polymer together. This causes the cell wall to lose strength and the bacteria to burst.

Antibiotic drug design is facilitated when an enzyme that is essential to the pathogen's survival is absent or very different in humans. In the example above, humans do not make peptidoglycan, therefore inhibitors of this process are selectively toxic to bacteria. Selective toxicity is also produced in antibiotics by exploiting differences in the structure of the ribosomes in bacteria, or how they make fatty acids.

Metabolic Control

Enzyme inhibitors are also important in metabolic control. Many metabolic pathways in the cell are inhibited by metabolites that control enzyme activity through allosteric regulation or substrate inhibition. A good example is the allosteric regulation of the glycolytic pathway. This catabolic pathway consumes glucose and produces ATP, NADH and pyruvate. A key step for the regulation of glycolysis is an early reaction in the pathway catalysed by phosphofructokinase-1 (PFK1). When ATP levels rise, ATP binds an allosteric site in PFK1 to decrease the rate of the enzyme reaction; glycolysis is inhibited and ATP production falls. This negative feedback control helps maintain a steady concentration of ATP in the cell. However, metabolic pathways are not just regulated through inhibition since enzyme activation is equally important. With respect to PFK1, fructose 2,6-bisphosphate and ADP are examples of metabolites that are allosteric activators.

Physiological enzyme inhibition can also be produced by specific protein inhibitors. This mechanism occurs in the pancreas, which synthesises many digestive precursor enzymes known as zymogens. Many of these are activated by the trypsin protease, so it is important to inhibit the activity of trypsin in the pancreas to prevent the organ from digesting itself. One way in which the activity of trypsin is controlled is the production of a specific and potent trypsin inhibitor protein in the pancreas. This inhibitor binds tightly to trypsin, preventing the trypsin activity that would otherwise be detrimental to the organ. Although the trypsin inhibitor is a protein, it avoids being hydrolysed as a substrate by the protease by excluding water from trypsin's active site and destabilising the transition state. Other examples of physiological enzyme inhibitor proteins include the barstar inhibitor of the bacterial ribonuclease barnase.

Pesticides

Many pesticides are enzyme inhibitors. Acetylcholinesterase (AChE) is an enzyme found in animals from insects to humans. It is essential to nerve cell function through its mechanism of breaking down the neurotransmitter acetylcholine into its constituents, acetate and choline. This is somewhat unusual among neurotransmitters as most, including serotonin, dopamine, and norepinephrine, are absorbed from the synaptic cleft rather than cleaved. A large number of AChE inhibitors are used in both medicine and agriculture. Reversible competitive inhibitors, such as edrophonium, physostigmine, and neostigmine, are used in the treatment of myasthenia gravis and in anaesthesia. The carbamate pesticides are also examples of reversible AChE inhibitors. The organophosphate pesticides such as malathion, parathion, and chlorpyrifos irreversibly inhibit acetylcholinesterase.

The herbicide glyphosate is an inhibitor of 3-phosphoshikimate 1-carboxyvinyltransferase, other herbicides, such as the sulfonylureas inhibit the enzyme acetolactate synthase. Both these enzymes are needed for plants to make branched-chain amino acids. Many other enzymes are inhibited by herbicides, including enzymes needed for the biosynthesis of lipids and carotenoids and the processes of photosynthesis and oxidative phosphorylation.

Natural Poisons

To discourage seed predators, pulses contain trypsin inhibitors that interfere with digestion.

Animals and plants have evolved to synthesise a vast array of poisonous products including secondary metabolites, peptides and proteins that can act as inhibitors. Natural toxins are usually small organic molecules and are so diverse that there are probably natural inhibitors for most metabolic processes. The metabolic processes targeted by natural poisons encompass more than enzymes in metabolic pathways and can also include the inhibition of receptor, channel and structural protein functions in a cell. For example, paclitaxel (taxol), an organic molecule found in the Pacific yew tree, binds tightly to tubulin dimers and inhibits their assembly into microtubules in the cytoskeleton.

Many natural poisons act as neurotoxins that can cause paralysis leading to death and have functions for defence against predators or in hunting and capturing prey. Some of these natural inhibitors, despite their toxic attributes, are valuable for therapeutic uses at lower doses. An example of a neurotoxin are the glycoalkaloids, from the plant species in the *Solanaceae* family (includes potato, tomato and eggplant), that are acetylcholinesterase inhibitors. Inhibition of this enzyme causes an uncontrolled increase in the acetylcholine neurotransmitter, muscular paralysis and then death. Neurotoxicity can also result from the inhibition of receptors; for example, atropine from deadly nightshade (*Atropa belladonna*) that functions as a competitive antagonist of the muscarinic acetylcholine receptors.

Although many natural toxins are secondary metabolites, these poisons also include peptides and proteins. An example of a toxic peptide is alpha-amanitin, which is found in relatives of the death cap mushroom. This is a potent enzyme inhibitor, in this case preventing the RNA polymerase II enzyme from transcribing DNA. The algal toxin microcystin is also a peptide and is an inhibitor of protein phosphatases. This toxin can contaminate water supplies after algal blooms and is a known carcinogen that can also cause acute liver hemorrhage and death at higher doses.

Proteins can also be natural poisons or antinutrients, such as the trypsin inhibitors that are found in some legumes, in the figure above. A less common class of toxins are toxic enzymes: these act as irreversible inhibitors of their target enzymes and work by chemically modifying their substrate enzymes. An example is ricin, an extremely potent protein toxin found in castor oil beans. This enzyme is a glycosidase that inactivates ribosomes. Since ricin is a catalytic irreversible inhibitor, this allows just a single molecule of ricin to kill a cell.

COMPETITIVE INHIBITION

Competitive inhibition is interruption of a chemical pathway owing to one chemical substance inhibiting the effect of another by competing with it for binding or bonding. Any metabolic or chemical messenger system can potentially be affected by this principle, but several classes of competitive inhibition are especially important in biochemistry and medicine, including the competitive form of enzyme inhibition, the competitive form of receptor antagonism, the competitive form of antimetabolite activity, and the competitive form of poisoning (which can include any of the aforementioned types).

In competitive inhibition of enzyme catalysis, binding of an inhibitor prevents binding of the target molecule of the enzyme, also known as the substrate. This is accomplished by blocking the binding site of the substrate – the active site – by some means. The V_{max} indicates the maximum velocity of the reaction, while the K_m is the amount of substrate needed to reach half of the V_{max}. K_m also plays a part in indicating the tendency of the substrate to bind the enzyme. Competitive inhibition can be overcome by adding more substrate to the reaction, which increases the chances of the enzyme and substrate binding. As a result, competitive inhibition alters only the K_m, leaving the V_{max} the same. This can be demonstrated using enzyme kinetics plots such as the Michaelis-Menten or the Lineweaver-Burk plot. Once the inhibitor is bound to the enzyme, the slope will be affected, as the K_m either increases or decreases from the original K_m of the reaction.

Most competitive inhibitors function by binding reversibly to the active site of the enzyme. As a result, many sources state that this is the defining feature of competitive inhibitors. This, however, is a misleading oversimplification, as there are many possible mechanisms by which an enzyme may bind either the inhibitor or the substrate but never both at the same time. For example, allosteric inhibitors may display competitive, non-competitive, or uncompetitive inhibition.

Mechanism

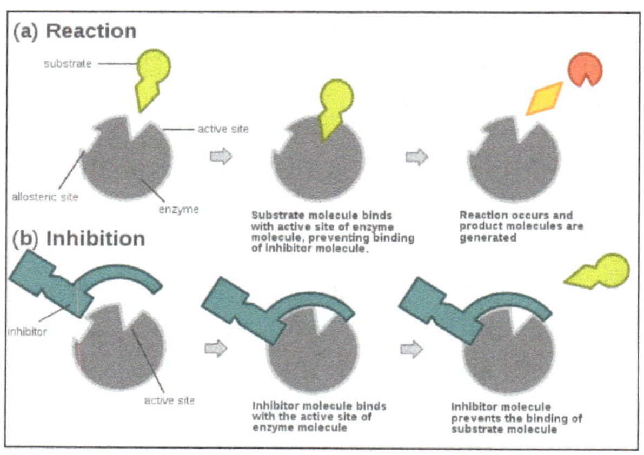

Diagram showing competitive inhibition.

In competitive inhibition, an inhibitor that resembles the normal substrate binds to the enzyme, usually at the active site, and prevents the substrate from binding. At any given moment, the enzyme may be bound to the inhibitor, the substrate, or neither, but it cannot bind both at the same time. During competitive inhibition, the inhibitor and substrate compete for the active site. The active site is a region on an enzyme which a particular protein or substrate can bind to. The active site will only allow one of the two complexes to bind to the site therefore either allowing for a reaction to occur or yielding it. In competitive inhibition the inhibitor resembles the substrate therefore taking its place and binding to the active site of an enzyme. Increasing the substrate concentration would diminish the "competition" for the substrate to properly bind to the active site and allow a reaction to occur. When the substrate is of higher concentration than that of the competitive inhibitor, it is more likely that the substrate will come into contact with the enzyme's active site than the inhibitor.

Competitive inhibitors are commonly used to make pharmaceuticals. For example, methotrexate is a chemotherapy drug that acts as a competitive inhibitor. It is structurally similar to the coenzyme, folate, which binds to the enzyme dihydrofolate reductase. This enzyme is part of the synthesis of DNA and RNA, and when methotrexate binds the enzyme, it renders it inactive, so it cannot synthesize DNA and RNA. Thus, the cancer cells are unable to grow and divide. Another example involves prostaglandins which are made in large amounts as a response to pain and can cause inflammation. Essential fatty acids form the prostaglandins and when this was discovered, it turned out that these were actually very good inhibitors to prostaglandins. These fatty acids inhibitors have been used as drugs to relieve pain because they can act as the substrate, and bind to the enzyme, and block prostaglandins.

An example of non-drug related competitive inhibition is in the prevention of browning of fruits and vegetables. For example, tyrosinase, an enzyme within mushrooms, normally binds to the substrate, monophenols, and forms brown o-quinones. Competitive substrates, such as 4-substituted benzaldehydes for mushrooms, compete with the substrate lowering the amount of the monophenols that bind. These inhibitory compounds added to the produce keep it fresh for longer periods of time by decreasing the binding of the monophenols that cause browning. This allows for an increase in produce quality as well as shelf life.

Competitive inhibition can be reversible or irreversible. If it is reversible inhibition, then effects of the inhibitor can be overcome by increasing substrate concentration. If it is irreversible, the only way to overcome it is to produce more of the target (and typically degrade and/or excrete the irreversibly inhibited target).

In virtually every case, competitive inhibitors bind in the same binding site (active site) as the substrate, but same-site binding is not a requirement. A competitive inhibitor could bind to an allosteric site of the free enzyme and prevent substrate binding, as long as it does not bind to the allosteric site when the substrate is bound. For example, strychnine acts as an allosteric inhibitor of the glycine receptor in the mammalian spinal cord and brain stem. Glycine is a major post-synaptic inhibitory neurotransmitter with a specific receptor site. Strychnine binds to an alternate site that reduces the affinity of the glycine receptor for glycine, resulting in convulsions due to lessened inhibition by the glycine.

In competitive inhibition, the maximum velocity (V_{max}) of the reaction is unchanged, while the apparent affinity of the substrate to the binding site is decreased (the K_d dissociation constant is apparently increased). The change in K_m (Michaelis-Menten constant) is parallel to the alteration in K_d, as one increases the other must decrease. When a competitive inhibitor is bound to an enzyme the K_m increases. This means the binding affinity for the enzyme is decreased, but it can be overcome by increasing the concentration of the substrate. Any given competitive inhibitor concentration can be overcome by increasing the substrate concentration. In that case, the substrate will reduce the availability for an inhibitor to bind, and, thus, outcompete the inhibitor in binding to the enzyme.

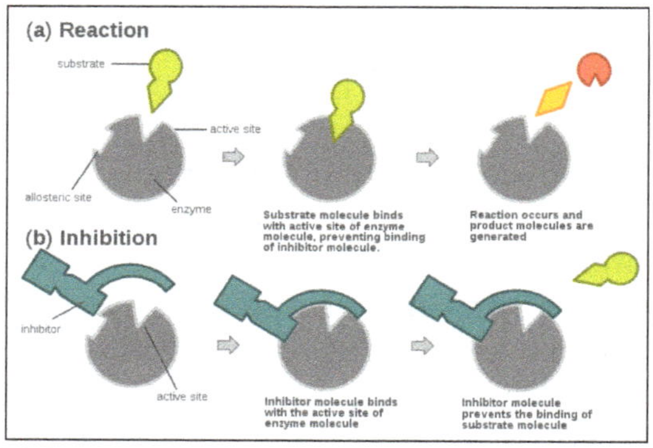

Competitive inhibition can also be allosteric, as long as the inhibitor and the substrate cannot bind the enzyme at the same time.

Biological Examples

After an accidental ingestion of a contaminated opioid drug desmethylprodine, the neurotoxic effect of 1-methyl-4-phenyl-1,2,3,6-tetrahydropyridine (MPTP) was discovered. MPTP is able to cross the blood brain barrier and enter acidic lysosomes. MPTP is biologically activated by MAO-B, an isozyme of monoamine oxidase (MAO) which is mainly concentrated in neurological disorders and diseases. Later, it was discovered that MPTP causes symptoms similar to that of Parkinson's disease. Cells in the central nervous system (astrocytes) include MAO-B that oxidizes MPTP to 1-methyl-4-phenylpyridinium (MPP+), which is toxic. MPP+ eventually travels to the extracellular fluid by a dopamine transporter, which ultimately causes the Parkinson's symptoms. However, competitive inhibition of the MAO-B enzyme or the dopamine transporter protects against the oxidation of MPTP to MPP+. A few compounds have been tested for their ability to inhibit oxidation of MPTP to MPP+ including methylene blue, 5-nitroindazole, norharman, 9-methylnorharman, and menadione. These demonstrated a reduction of neurotoxicity produced by MPTP.

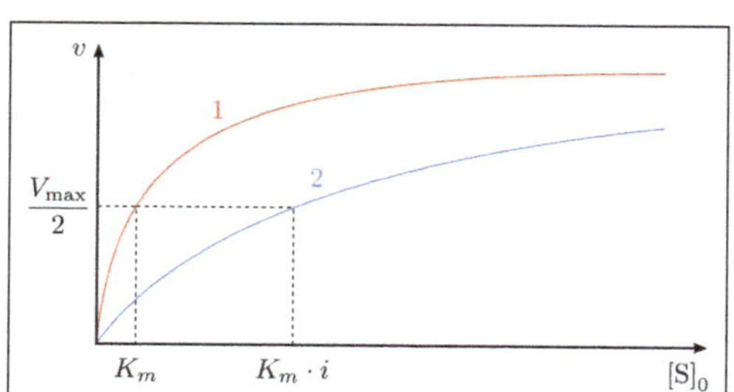

Michaelis-Menten plot of the reaction velocity (v) against substrate concentration [S] of normal enzyme activity (1) compared to enzyme activity with a competitive inhibitor

(2) Adding a competitive inhibitor to an enzymatic reaction increases the K_m of the reaction, but the V_{max} remains the same.

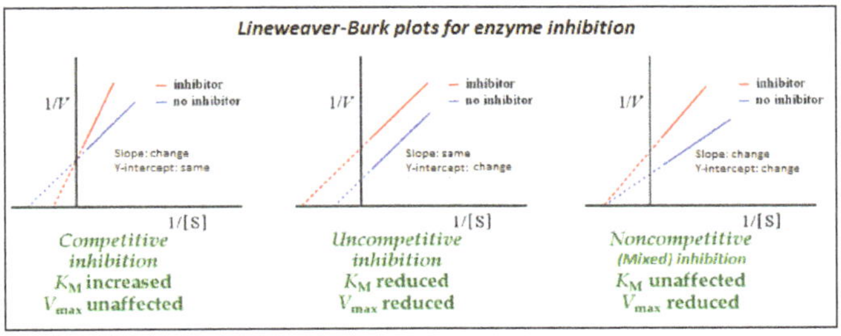

Lineweaver-Burk plot, the reciprocal of the Michaelis-Menten plot, of the reciprocal of velocity (1/V) vs the reciprocal of the substrate concentration (1/[S]) of normal enzyme activity (blue) compared to enzyme activity with a competitive inhibitor (red). Adding a competitive inhibitor to an enzymatic reaction increases the K_m of the reaction, but the V_{max} remains the same.

Sulfa drugs also act as competitive inhibitors. For example, sulfanilamide competitively binds to the enzyme in the dihydropteroate synthase (DHPS) active site by mimicking the substrate para-aminobenzoic acid (PABA). This prevents the substrate itself from binding which halts the production of folic acid, an essential nutrient. Bacteria must synthesize folic acid because they do not have a transporter for it. Without folic acid, bacteria cannot grow and divide. Therefore, because of sulfa drugs' competitive inhibition, they are excellent antibacterial agents. An example of competitive inhibition was demonstrated experimentally for the enzyme succinic dehydrogenase, which catalyzes the oxidation of succinate to fumarate in the Krebs cycle. Malonate is a competitive inhibitor of succinic dehydrogenase. The binding of succinic dehydrogenase to the substrate, succinate, is competitively inhibited. This happens because malonate's chemistry is similar to succinate. Malonate's ability to inhibit binding of the enzyme and substrate is based on the ratio of malonate to succinate. Malonate binds to the active site of succinic dehydrogenase so that succinate cannot. Thus, it inhibits the reaction.

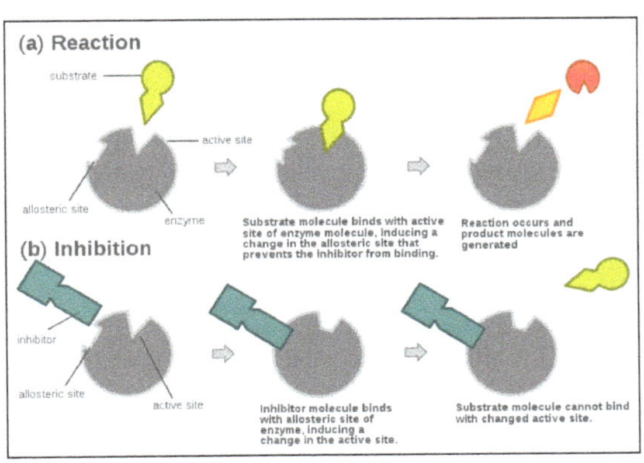

Another possible mechanism for allosteric competitive inhibition.

Equation

The Michaelis-Menten Model can be an invaluable tool to understanding enzyme kinetics. According to this model, a plot of the reaction velocity (V_0) associated with the concentration [S] of the substrate can then be used to determine values such as V_{max}, initial velocity, and K_m ($V_{max}/2$ or affinity of enzyme to substrate complex).

Competitive inhibition increases the apparent value of the Michaelis-Menten constant, K_m^{app}, such that initial rate of reaction, V_0, is given by,

$$V_0 = \frac{V_{max}[S]}{K_m^{app} + [S]}$$

where $K_m^{app} = K_m(1+[I]/K_i)$, K_i is the inhibitor's dissociation constant and $[I]$ is the inhibitor concentration.

V_{max} remains the same because the presence of the inhibitor can be overcome by higher substrate concentrations. K_m^{app}, the substrate concentration that is needed to reach $V_{max}/2$, increases with the presence of a competitive inhibitor. This is because the concentration of substrate needed to reach V_{max} with an inhibitor is greater than the concentration of substrate needed to reach V_{max} without an inhibitor.

Derivation

In the simplest case of a single-substrate enzyme obeying Michaelis-Menten kinetics, the typical scheme,

$$\mathrm{E} + \mathrm{S} \underset{k_{-1}}{\overset{k_1}{\rightleftharpoons}} \mathrm{ES} \overset{k_2}{\rightarrow} \mathrm{E} + \mathrm{P}$$

is modified to include binding of the inhibitor to the free enzyme:

$$\mathrm{EI} + \mathrm{S} \underset{k_3}{\overset{k_{-3}}{\rightleftharpoons}} \mathrm{E} + \mathrm{S} + \mathrm{I} \underset{k_{-1}}{\overset{k_1}{\rightleftharpoons}} \mathrm{ES} + \mathrm{I} \overset{k_2}{\rightarrow} \mathrm{E} + \mathrm{P} + \mathrm{I}$$

The inhibitor does not bind to the ES complex and the substrate does not bind to the EI complex. It is generally assumed that this behavior is indicative of both compounds binding at the same site, but that is not strictly necessary. As with the derivation of the Michaelis-Menten equation, assume that the system is at steady-state, i.e. the concentration of each of the enzyme species is not changing:

$$\frac{d[\mathrm{E}]}{dt} = \frac{d[\mathrm{ES}]}{dt} = \frac{d[\mathrm{EI}]}{dt} = 0.$$

Furthermore, the known total enzyme concentration is $[\mathrm{E}]_0 = [\mathrm{E}] + [\mathrm{ES}] + [\mathrm{EI}]$, and the velocity is measured under conditions in which the substrate and inhibitor concentrations do not change substantially and an insignificant amount of product has accumulated.

We can therefore set up a system of equations:

$$[\mathrm{E}]_0 = [\mathrm{E}] + [\mathrm{ES}] + [\mathrm{EI}]$$

$$\frac{d[\mathrm{E}]}{dt} = 0 = -k_1[\mathrm{E}][\mathrm{S}] + k_{-1}[\mathrm{ES}] + k_2[\mathrm{ES}] - k_3[\mathrm{E}][\mathrm{I}] + k_{-3}[\mathrm{EI}]$$

$$\frac{d[\mathrm{ES}]}{dt} = 0 = k_1[\mathrm{E}][\mathrm{S}] - k_{-1}[\mathrm{ES}] - k_2[\mathrm{ES}]$$

$$\frac{d[\mathrm{EI}]}{dt} = 0 = k_3[\mathrm{E}][\mathrm{I}] - k_{-3}[\mathrm{EI}]$$

where $[\mathrm{S}]$, $[\mathrm{I}]$ and $[\mathrm{E}]_0$ are known. The initial velocity is defined as $V_0 = d[\mathrm{P}]/dt = k_2[\mathrm{ES}]$, so we need to define the unknown $[\mathrm{ES}]$ in terms of the knowns $[\mathrm{S}]$, $[\mathrm{I}]$ and $[\mathrm{E}]_0$.

From equation $\frac{d[\mathrm{ES}]}{dt} = 0 = k_1[\mathrm{E}][\mathrm{S}] - k_{-1}[\mathrm{ES}] - k_2[\mathrm{ES}]$, we can define E in terms of ES by rearranging to,

$$k_1[\mathrm{E}][\mathrm{S}] = (k_{-1} + k_2)[\mathrm{ES}]$$

Dividing by $k_1[\mathrm{S}]$ gives,

$$[\mathrm{E}] = \frac{(k_{-1} + k_2)[\mathrm{ES}]}{k_1[\mathrm{S}]}$$

As in the derivation of the Michaelis-Menten equation, the term $(k_{-1} + k_2)/k_1$ can be replaced by the macroscopic rate constant K_m:

$$[\mathrm{E}] = \frac{K_m[\mathrm{ES}]}{[\mathrm{S}]}$$

Substituting equation $[\mathrm{E}] = \frac{K_m[\mathrm{ES}]}{[\mathrm{S}]}$ into equation $\frac{d[\mathrm{EI}]}{dt} = 0 = k_3[\mathrm{E}][\mathrm{I}] - k_{-3}[\mathrm{EI}]$, we have

$$0 = \frac{k_3[\mathrm{I}]K_m[\mathrm{ES}]}{[\mathrm{S}]} - k_{-3}[\mathrm{EI}]$$

Rearranging, we find that,

$$[\mathrm{EI}] = \frac{K_m k_3[\mathrm{I}][\mathrm{ES}]}{k_{-3}[\mathrm{S}]}$$

At this point, we can define the dissociation constant for the inhibitor as $K_i = k_{-3} / k_3$, giving:

$$[\mathrm{EI}] = \frac{K_m[\mathrm{I}][\mathrm{ES}]}{K_i[\mathrm{S}]}$$

At this point, substitute equation $[\mathrm{E}] = \frac{K_m[\mathrm{ES}]}{[\mathrm{S}]}$ and equation $[\mathrm{EI}] = \frac{K_m[\mathrm{I}][\mathrm{ES}]}{K_i[\mathrm{S}]}$ into equation $[\mathrm{E}]_0 = [\mathrm{E}] + [\mathrm{ES}] + [\mathrm{EI}]$:

$$[\mathrm{E}]_0 = \frac{K_m[\mathrm{ES}]}{[\mathrm{S}]} + [\mathrm{ES}] + \frac{K_m[\mathrm{I}]\,[\mathrm{ES}]}{K_i[\mathrm{S}]}$$

Rearranging to solve for ES, we find

$$[\mathrm{E}]_0 = [\mathrm{ES}]\left(\frac{K_m}{[\mathrm{S}]} + 1 + \frac{K_m[\mathrm{I}]}{K_i[\mathrm{S}]}\right) = [\mathrm{ES}]\frac{K_m K_i + K_i[\mathrm{S}] + K_m[\mathrm{I}]}{K_i[\mathrm{S}]}$$

$$[\mathrm{ES}] = \frac{K_i[\mathrm{S}][\mathrm{E}]_0}{K_m K_i + K_i[\mathrm{S}] + K_m[\mathrm{I}]}$$

Returning to our expression for V_0 , we now have:

$$V_0 = k_2[\mathrm{ES}] = \frac{k_2 K_i[\mathrm{S}][\mathrm{E}]_0}{K_m K_i + K_i[\mathrm{S}] + K_m[\mathrm{I}]}$$

$$V_0 = \frac{k_2[\mathrm{E}]_0[\mathrm{S}]}{K_m + [\mathrm{S}] + K_m\frac{[\mathrm{I}]}{K_i}}$$

Since the velocity is maximal when all the enzyme is bound as the enzyme-substrate complex, $V_{\max} = k_2[\mathrm{E}]_0$. Replacing and combining terms finally yields the conventional form:

$$V_0 = \frac{V_{\max}[\mathrm{S}]}{K_m\left(1 + \frac{[\mathrm{I}]}{K_i}\right) + [\mathrm{S}]}$$

To compute the concentration of competitive inhibitor [I] that yields a fraction f_{V_0} of velocity V_0 where:

$$[\mathrm{I}] = \left(\frac{1}{fv_0} - 1\right)K_i\left(1 + \frac{[\mathrm{S}]}{K_m}\right)$$

UNCOMPETITIVE INHIBITOR

Uncompetitive inhibition, also known as anti-competitive inhibition, takes place when an enzyme inhibitor binds only to the complex formed between the enzyme and the substrate (the E-S complex). Uncompetitive inhibition typically occurs in reactions with two or more substrates or products.

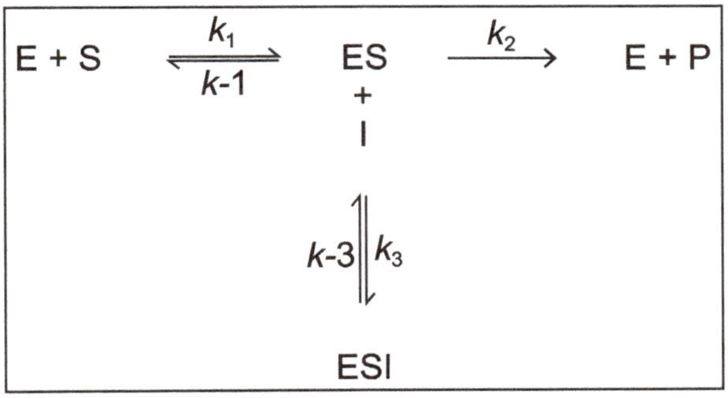

General representation of uncompetitive inhibition.

While uncompetitive inhibition requires that an enzyme-substrate complex must be formed, non-competitive inhibition can occur with or without the substrate present.

Uncompetitive inhibition is distinguished from competitive inhibition by two observations: first uncompetitive inhibition cannot be reversed by increasing [S] and second, the Lineweaver–Burk plot yields parallel rather than intersecting lines. This behavior is found in the inhibition of acetylcholinesterase by tertiary amines (R_3N). Such compounds bind to the enzyme in its various forms, but the acyl-intermediate-amine complex cannot break down into enzyme plus product.

Mechanism

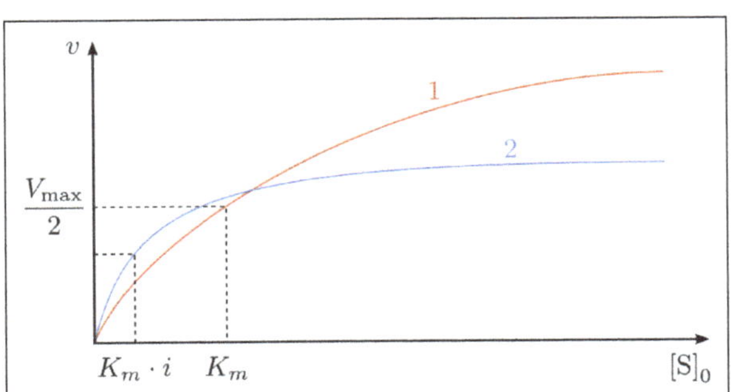

Michaelis-Menten plot of uncompetitive inhibition.

Uncompetitive inhibition is unique in that the inhibitor binds to the enzyme-substrate complex. This could imply that the binding site for the inhibitor is accessible only after the enzyme has bound to its substrate.

As inhibitor binds, the amount of ES complex is reduced. This reduction in the effective concentration of the ES complex can be explained by the fact that having the inhibitor bound to the ES complex essentially converts it to ESI complex, which is considered a separate complex altogether. This reduction in ES complex decreases the maximum enzyme activity (V_{max}), as it takes longer for the substrate or product to leave the active site. The reduction in K_m - the substrate concentration at which the enzyme can operate at half of its maximal velocity, often used to approximate an enzyme's affinity for a substrate - can also be linked back to the decrease in ES complex. Le Chatelier's principle opposes this decrease and attempts to make up for the loss of ES, so more free enzyme is converted to the ES form, and the amount of ES increases overall. An increase in ES generally indicates that the enzyme has a high degree of affinity for its substrate. K_m decreases as affinity for a substrate increases, though it is not a perfect predictor of affinity since it accounts for other factors as well; regardless, this increase in affinity will be accompanied by a decrease in K_m.

In general, uncompetitive inhibition works best when substrate concentration is high. An uncompetitive inhibitor need not resemble the substrate of the reaction it is inhibiting.

Mathematical Definition

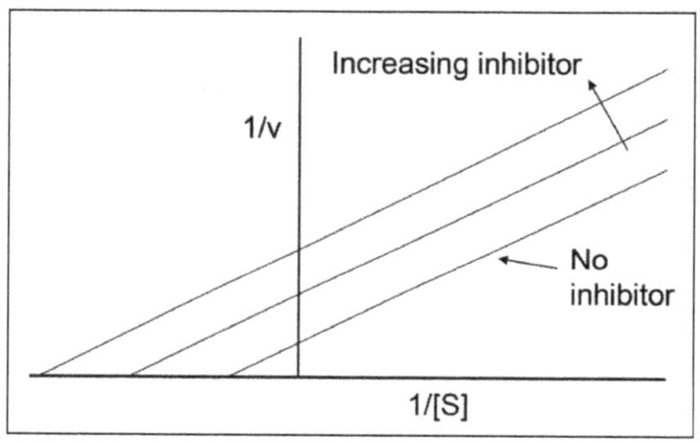

Lineweaver–Burk plot of uncompetitive enzyme inhibition.

The Lineweaver–Burk equation states that:

$$\frac{1}{v} = \frac{K_m}{V_{max}[S]} + \frac{1}{V_{\max}}$$

Where v is the initial reaction velocity, K_m is the Michaelis–Menten constant, $V_{\max}$ is the maximum reaction velocity, and $[S]$ is the concentration of the substrate.

The Lineweaver–Burk plot for an uncompetitive inhibitor produces a line parallel to the original enzyme-substrate plot, but with a higher y-intercept, due to the presence of an inhibition term $\frac{[I]}{K_i}$:

$$\frac{1}{v} = \frac{K_m}{V_{max}[S]} + \frac{1 + \frac{[I]}{K_i}}{V_{max}}$$

Where $[I]$ is the concentration of the inhibitor and K_i is an inhibition constant characteristic of the inhibitor.

The Michaelis-Menten equation is altered to:

$$V_0 = \frac{V_{max}[S]}{K_m + \alpha'[S]}$$

where,

$$\alpha' = 1 + \frac{[I]}{K_I'} \text{ and } K_I' = \frac{[ES][I]}{[ESI]}$$

As described by the equation above, at high concentrations of substrate, V_o approaches V_{max}/α'. Thus, an uncompetitive inhibitor lowers the measured V_{max}. Apparent K_m also decreases, because [S] required to reach one-half V_{max} decreases by the factor α'. It is important to note that V_{max} and K_m decrease at the same rate as a result of the inhibitor. This is apparent when viewing a Lineweaver-Burk plot of uncompetitive enzyme inhibition: the ratio between V and K_m remains the same with or without an inhibitor present.

Implications and uses in Biological Systems

The unique traits of uncompetitive inhibition lead to a variety of implications for the inhibition's effects within biological and biochemical systems. Uncompetitive inhibition is present within biological systems in a number of ways. In fact, it often becomes clear that the traits of inhibition specific to uncompetitive inhibitors, such as their tendency to act at their best at high concentrations of substrate, are essential to some important bodily functions operating properly.

Involvement in Cancer Mechanisms

Uncompetitive mechanisms are involved with certain types of cancer. Human alkaline phosphatases such as CGAP have been found to be over-expressed in certain types of cancers, and those phosphotases often operate via uncompetitive inhibition. It has also been found that a number of the genes that code for human alkaline phosphatases (TSAPs) are inhibited uncompetitively by amino acids such as Leucine and Phenylalanine. Studies of the involved amino acid residues have been undertaken in attempts to regulate alkaline phosphatase activity and learn more about said activity's relevance to cancer.

Additionally, uncompetitive inhibition works alongside TP53 to help repress the activity of cancer cells and prevent tumorigenesis in certain forms of the illness, as it inhibits G6PD (glucose-6-phosphate dehydrogenase, an enzyme primarily involved in certain metabolic pathways). One of the side roles G6PD is responsible for is helping to regulate is the control of reactive oxygen levels, as reactive oxygen species must be kept at appropriate levels to allow cells to survive. When G6PD's substrate concentration is high, uncompetitive inhibition of the enzyme becomes far more effective. As substrate concentration increases, the amount of ES complex increases as well, and with more ES complex to bind, uncompetitive inhibitors become far more active. This inhibition works such that the higher the concentration of substrate is in the system initially, the harder it is to reach the maximum velocity of the reaction. At low initial substrate concentrations, increasing the concentration of substrate is sometimes enough to entirely or even fully restore the enzyme's function, but as soon as initial concentration increases past a certain point, reaching the maximal enzyme velocity is all but impossible. This extreme sensitivity to substrate concentration within the cancer mechanism implicates uncompetitive inhibition rather than mixed inhibition, which displays similar traits but is often less sensitive to substrate concentration due to some inhibitor binding to free enzymes regardless of the substrate's presence. As such, the extreme strength of uncompetitive inhibitors at high substrate concentrations and the overall sensitivity to substrate amount indicates that only uncompetitive inhibition can make this type of process possible.

Importance in Cell and Organelle Membranes

Although this form of inhibition is present in various diseases within biological systems, it does not necessarily only relate to pathologies. It can be involved in typical bodily functions. For example, active sites capable of uncompetitive inhibition appear to be present in membranes, as removing lipids from cell membranes and making active sites more accessible through conformational changes has been shown to invoke elements resembling the effects of uncompetitive inhibition (i.e. both K_M and V_{Max} decrease). In mitochondrial membrane lipids specifically, removing lipids decreases the alpha-helix content in mitochondria and leads to changes in ATPase resembling uncompetitive inhibition.

This presence of uncompetitive enzymes in membranes has also been supported in a number of other studies. A type of protein called an Arf protein involved in regulating membrane activity was being studied, and it was found that an inhibitor called BFA trapped one of Arf's intermediates via uncompetitive inhibition. This made it clear that this type of inhibition exists within various types of cells and organelles as opposed to just in pathological cells. In fact, BFA was found to relate to the activity of the Golgi apparatus and its role in regulating movement across the cell membrane.

Presence in the Cerebellar Granule Layer

Uncompetitive inhibition can play roles in various other parts of the body as well. It is part of the mechanism by which NMDA (N-methyl-D-aspartate) glutamate receptors are inhibited in the brain, for example. Specifically, this type of inhibition impacts the granule cells that make up a layer of the cerebellum. These cells have the aforementioned NMDA receptors, and the activity of said receptors typically increases as ethanol is consumed. This often leads to withdrawal symptoms if said ethanol is removed. Various uncompetitive blockers act as antagonists at the receptors and modify the process, with one example being an inhibitor called memantine. In fact, in similar cases (involving over-expression of NMDA, though not necessarily via ethanol), it has been shown that uncompetitive inhibition helps in nullifying the over-expression due to its particular properties. Since uncompetitive inhibitors block high concentrations of substrates very efficiently, their traits alongside the innate characteristics of the receptors themselves lead to very effective blocking of NMDA channels when they are excessively open due to massive amounts of NMDA agonists.

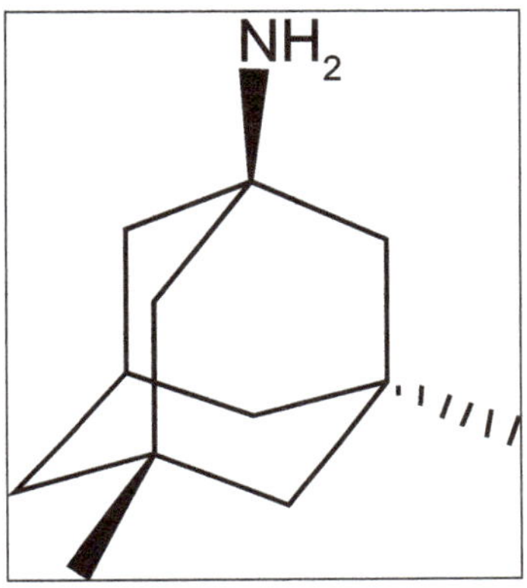

NMDA Inhibitor Memantine.

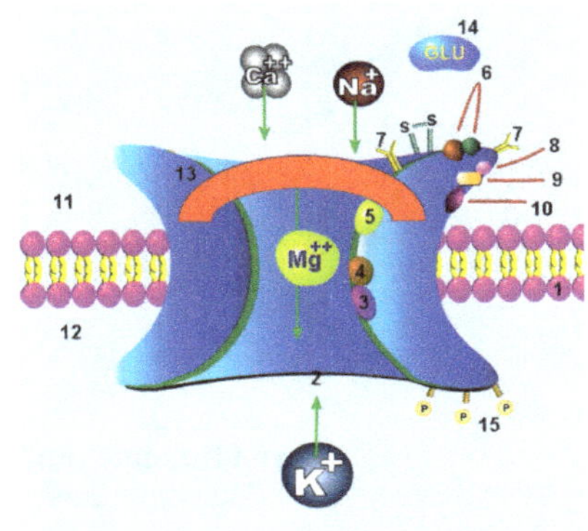

Inhibited NMDA Receptor. Substrate is bound and the active site is blocked by the (red) inhibitor.

TARGETED COVALENT INHIBITORS

Targeted covalent inhibitors (TCIs) or Targeted covalent drugs are rationally designed inhibitors that bind and then bond to their target proteins. These inhibitors possess a bond-forming functional group of low chemical reactivity that, following binding to the target protein, is positioned to react rapidly with a specific nucleophilic residue at the target site and form a bond.

This illustration describes the mechanism by which covalent drugs irreversibly bind and modify the protein effectively silencing its activity.

Over the last 100 years covalent drugs have made a major impact on human health and have been highly successful drugs for the pharmaceutical industry. These inhibitors react with their target proteins to form a covalent complex in which the protein has lost its function. The majority of these successful drugs, which include penicillin, omeprazole, clopidogrel, and aspirin were discovered through serendipity in phenotypic screens.

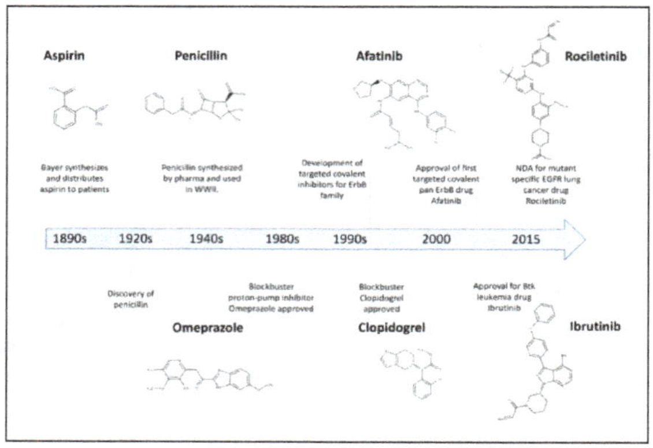

A brief timeline representation of the history of covalent drugs that have been approved to be marketed.

For the past 20 years key changes in screening approaches, along with safety concerns, have made pharma reluctant to pursue covalent inhibitors in a systematic way.

Recently, there has been considerable attention to using rational drug design to create highly selective covalent inhibitors called targeted covalent inhibitors. The first published example of a targeted covalent drug was for the EGFR kinase. but this has now broadened to other kinases and other protein families.

Advantages of Covalent Drugs

Potency

Covalent bonding can lead to potencies and ligand efficiencies that are either exceptionally high or, for irreversible covalent interactions, even essentially infinite. Covalent bonding thus allows high potency to be routinely achieved in compounds of low molecular mass, along with all the beneficial pharmaceutical properties that are associated with small size.

Selectivity

Covalent inhibitors can be designed to target a nucleophile that is unique or rare across a protein family. thereby ensuring that covalent bond formation cannot occur with most other family members. This approach can lead to high selectivity against closely related proteins because although the inhibitor might bind transiently to the active sites of such proteins, it will not covalently label them if they lack the targeted nucleophilic residue in the appropriate position.

Pharmacodynamics

The restoration of pharmacological activity after covalent irreversible inhibition requires re-synthesis of the protein target. This has important and potentially advantageous consequences for drug pharmacodynamics in which the level and frequency of dosing relates to the extent and duration of the resulting pharmacological effect.

Built-in-biomarker

Covalent inhibitors can be used to assess target engagement which can sometimes be used pre-clinically and clinically to assess the relationship between dose of drug and efficacy or toxicity. This approach was used for covalent Btk inhibitors pre-clinically and clinically to understand the relationship between dose administered and efficacy in animal models of arthritis and target occupancy in a clinical study of healthy volunteers.

Design of Covalent Drugs

The design of covalent drugs requires careful optimization of both the non-covalent binding affinity (which is reflected in Ki) and the reactivity of the electrophilic warhead (which is reflected in k2).

$$E + I \underset{K_i}{\rightleftharpoons} E \bullet I \underset{k_{-2}}{\overset{k_2}{\rightleftharpoons}} E - I$$

Initial non-covalent complex (E • I); Final covalent complex (E - I)

Mechanism of Action of Covalent Drugs.

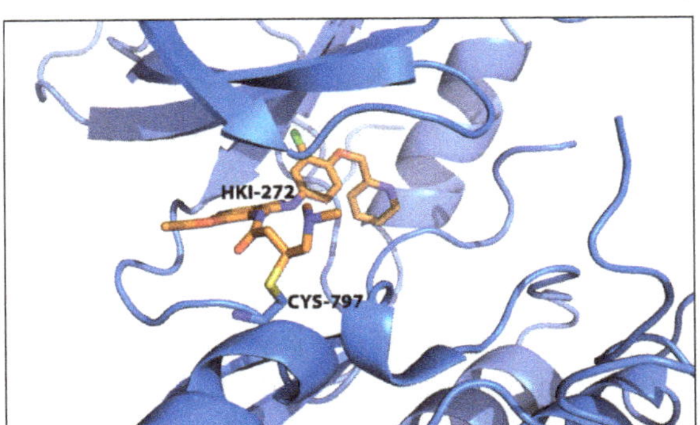

EGFR kinase T790M mutant covalently inhibited by HKI-272 (neratinib) at Cys-797 (PDB ID: 2JIV).

The initial design of TCIs involves three key steps. First, bioinformatics analysis is used to identify a nucleophilic amino acid (for example, cysteine) that is either inside or near to a functionally relevant binding site on a drug target, but is rare in that protein family.

Next, a reversible inhibitor is identified for which the binding mode is known. Finally, structure-based computational methods are used to guide the design of modified ligands that have electrophilic functionality, and are positioned to react specifically with the nucleophilic amino acid in the target protein.

Toxicity Risks Associated with Covalent Modification of Proteins

There has been a reluctance for modern drug discovery programs to consider covalent inhibitors due to toxicity concerns. An important contributor has been the drug toxicities of several high-profile drugs believed to be caused by metabolic activation of reversible drugs For example, high dose acetaminophen can lead to the formation of the reactive metabolite N-acetyl-p-benzoquinone imine. Also, covalent inhibitors such as beta lactam antibiotics which contain weak electrophiles can lead to idiosyncratic toxicities (IDT) in some patients. It has been noted that many approved covalent inhibitors have been used safely for decades with no observed idiosyncratic toxicity. Also, that IDTs are not limited to proteins with a covalent mechanism of action. A recent analysis has noted that the risk of idiosyncratic toxicities may be mitigated through lower doses of administered drug. Doses of less than 10 mg per day rarely lead to IDT irrespective of the drug mechanism.

TCIs in Clinical Development

Despite the apparent lack of attention towards covalent inhibitor drug discovery by most pharmaceutical companies, there are several examples of covalent drugs that have been approved or are progressing to late-stage clinical development.

KRAS and Lung, Colorectal Cancer

AMG 510 by Amgen is a KRAS p.G12C covalent inhibitor that has recently finished Phase I clinical trial. The drug elicited partial responses in half of evaluable patients with KRAS G12C-mutant non–small cell lung cancer, and led to stable disease in most evaluable patients with colorectal (or appendix) cancer.

EGFR and Lung Cancer

The second generation EGFR inhibitor Afatinib has been approved for the treatment of EGFR driven lung cancer and Dacomitinib is in late stage clinical testing. The third generation EGFR inhibitors which target mutant EGFR which is specific to the tumor but are selective against wild-type EGFR that are expected to lead to a wider therapeutic index.

ErbB Family and Breast Cancer

There has also been progress in the treatment of metastatic breast cancer with the irreversible inhibitor Neratinib which targets the ErbB family of tyrosine kinases.

Neratinib is in multiple Phase 2 and also a Phase 3 clinical trial for the treatment of breast cancer.

Btk and Leukemia

Ibrutinib, a covalent inhibitor of Bruton's tyrosine kinase, has been approved for the treatment of chronic lymphocytic leukemia, waldenstrom's macroglobulinemia and mantle cell lymphoma.

ZINC FINGER INHIBITOR

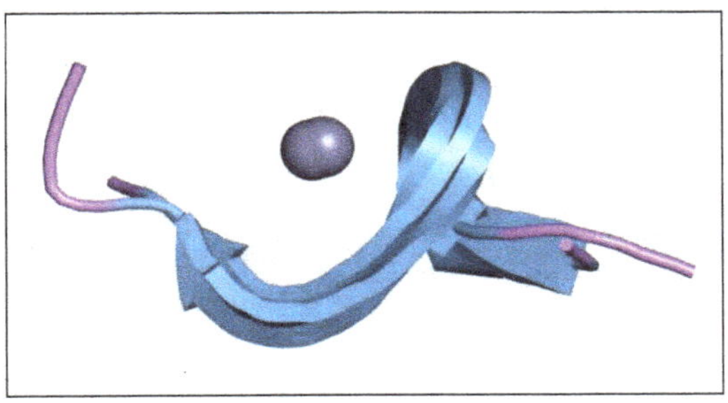

This image shows the typical structure of NCp7, which is targeted by zinc finger inhibitors when combating HIV.

Zinc finger inhibitors, or zinc ejectors, are substances or compounds that interact adversely with zinc fingers and cause them to release their zinc from its binding site, disrupting the conformation of the polypeptide chain and rendering the zinc fingers ineffective, thereby preventing them from performing their associated cellular functions. This is typically accomplished through chelation of the zinc binding site. As zinc fingers are known to be involved in m-RNA regulation, reverse transcription, protection of synthesized viral DNA, transcription inhibition, and initial integration processes, prevention of zinc finger function can have drastic effects on the function of the cell or virus.

Zinc finger inhibitors are typically used to combat HIV. HIV treatments usually rely on targeting reverse transcriptases and proteases. However, these methods are proving to be ineffective due to the development of resistant strains of the virus or due to the stoppage of the treatment. This method of using zinc finger inhibitors to target and destabilize zinc fingers represents a new method of fighting HIV. Other viruses such as SARS, polio, Ebola, measles, human coxsackie, Dengue, rabies, human hepatitis, human parainfluenza and human respiratory syncytical have similar zinc finger motifs and could potentially benefit from zinc finger inhibitor technology.

Zinc ejectors were patented in 2008 and some have entered Phase I/II trials as a HIV drug.

Zinc Finger Inhibitor Target: Nucleocapsid Protein

The HIV-1 nucleocapsid protein 7 (NCp7) is the protein targeted by zinc ejectors. NCp7 is initially formed as part of the gag polypeptide and follows a gag-knuckle zinc finger conformation. In its lifetime, NCp7 facilitates the unwinding of tRNA, acts as a primer for reverse transcription, chaperones nucleic acids within the capsid of HIV-1, helps integrate the viral RNA into budding virions, and is intimately involved in the replication of HIV-1 in both the early phase and late phase. These processes play critical roles in the replication of HIV-1 thus making NCp7 a prime target for drugs seeking to contravene the replication process.

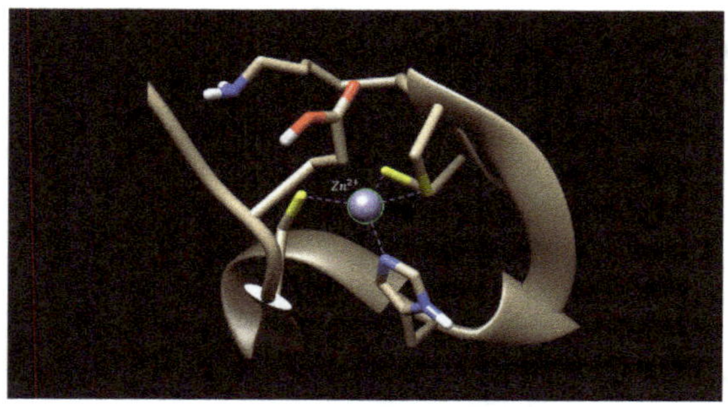

This image shows the zinc finger structure of the HIV-1 nucleocapsid protein. The cysteines of the CCHC motif are given in yellow; the histidine in blue. The dashed lines represent the coordinate bonds between the Zn ion and the respective residues.

NCp7 is a 55-amino acid protein that is highly basic and consists of two gag-knuckle motifs. These motifs contain two peptide units of Cys-X2-Cys-X4-His-X4-Cys (CCHC), where the X represents a substituted amino acid, that make up the zinc (II) ion binding sites. The binding of zinc (II) in the CCHC binding site is necessary for the domain to be functional and for the stabilization of the conformation of the structure, allowing the NCp7 to carry out the processes required for HIV replication. Since the CCHC binding site is mutation resistant and involved in the replication of HIV-1, it makes a prime candidate for the prevention of HIV through zinc ejectors. By inhibiting the function of NCp7, the viral replication is affected and a non-functional virus that is unable to infect its host is produced.

Zinc Ejector Compounds

Azodicarbonamide

Azodicarbonamide (ADA) was the first zinc ejector to go into clinical trial for treatment of HIV. ADA inhibits HIV by electophilically attacking the sulfur atoms of the zinc coordinated cysteine. This electrophilic interaction destabilizes the zinc binding site making it easier for the zinc ion to be withdrawn due to the new arrangement of bonds. The binding site then performs a disulfide exchange, forming new intermolecular disulfide bonds, and rearrangement occurs placing the zinc finger in a conformation that inhibits its function.

C-nitroso

3-nitrosobenzamide (NOBA) and 6-nitroso-1,2-benzopyrone (NOBP) were the first compounds to demonstrate an ability to inhibit infection of HIV by ejecting zinc from NCp7. In the same manner as ADA, the compounds interact with an 18-residue polypeptide on the N terminal zinc knuckle region of the HIV nucleocapsid protein which causes ejection of the zinc from the region by covalently modifying the cysteine residues. Studies suggest that NOBA and NOBP were able to inhibit HIV-1 infection by inhibiting reverse transcription without an apparent impairment of reverse transcriptase. This reiterates the role of NCp7 in reverse transcription.

2,2'-Di-Thiobisbenzamide (DIBA)

DIBAs act similarly to ADA, NOBA and NOBP. They react with the cysteine residues on the zinc finger of the NCp7 and cause a covalent conformation change which ejects the zinc from the zinc finger domain. Though DIBAs initially seemed to be promising antiviral candidates, there were clinical issues with their stability. DIBAs tend to cyclize into benzisothiazolones which do not have the same potency when used to combat retroviruses as the original compound. Additionally, glutathione can reduce the disulfide bonds in DIBA thereby restricting its function *in vitro*.

Mercaptobenzamides

Mercaptobenzamide prodrugs transfer acyl group to 36th cysteine residue of NCp7 and the acyl group then migrates to neighboring lysine residue that triggers the ejection of Zn. The mercaptobenzamide and its corresponding prodrug yielded additivity to synergy when combined with known HIV drugs. Moreover they showed no cytotoxicity.

Pyridinioalkanoyl Thiolesters (PATES)

N-[2-(5-pyridiniovaleroylthio)benzoyl] sulfacetamide bromide (referred to as compound 45) is a pyridinioalkanoyl thiolester that can function as a zinc ejector. Once activated with silver, compound 45 uses its pyridinioalkanoyl groups to covalently modify NCp7, specifically altering cysteines 36 and 49 on the carboxyl-terminal zinc finger. It ejects the zinc from the zinc binding sites in a two steps. The zinc in the carboxyl-terminal zinc finger is released first, followed by the ejection of zinc from the amino-terminal zinc finger.

Bis-Thiadizolbenzene-1,2-Diamine

Bis-Thiadizolbenzene-1,2-diamine (NV038) is one of the newer zinc ejectors. NV038 is found to effect the function of the zinc finger after the virus has entered the cell but before reverse transcription is completed. NV038, like other zinc ejector compounds, chelates zinc to remove it from its binding site. However, it is thought to act through a different mechanism than many of the other zinc ejectors due to its structural features.

Its structure would not readily allow thiol-disulfide interchange or acyl transfer to cysteine. Instead, NV038 is believed to react with the zinc using its two carbonyl oxygens found in the esters.

IRREVERSIBLE INHIBITORS

Irreversible inhibitors are covalently or noncovalently bound to the target enzyme and dissociates very slowly from the enzyme. There are three types of irreversible inhibitors: group-specific reagents, reactive substrate analogs also known as affinity labels and suicide inhibitors.

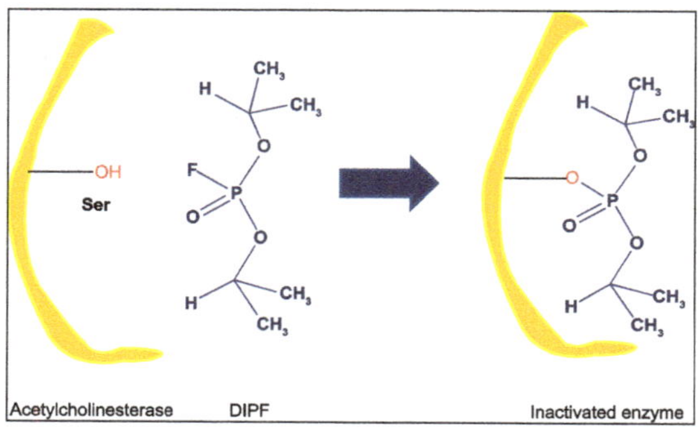

- Group specific reagents react with specific amino acid side chains like diisopropylphosphofluoridate (DIPF) and iodoacetamide. For example, only 1 of the 28 serine residues in chymotrypsin is modified by DIPF. This means that this specific residue is especially reactive; moreover, it is implied that this specific residue lies in the active site of the enzyme chymotrypsin. DIPF has also provided data that suggests, through its binding with active serine residues, that there is indeed a reactive serine residue contained within the active site of the enzyme Acetylcholinesterase. The inactivating functionality of DIPF and similarly-shaped molecules in acetylcholinesterase is representative of a group of compounds, known as nerve agents.

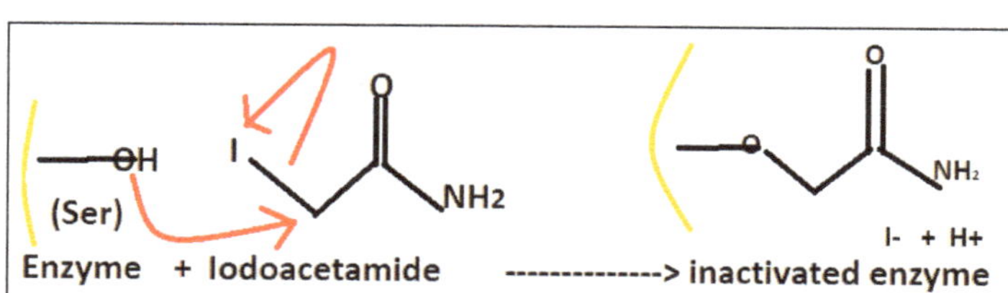

- Affinity labels (Reactive substrate analogs) are structurally similar to the substrate that can covalently bind to the active site and are therefore more specific than group specific reagents. An example is Tosyl-L-phenylalanine chloromethyl ketone

(TPCK) which is an analog for chymotrypsin which binds to the active site and reacts irreversibly with the histidine residue to inhibit the enzyme. Another example is triose phophate isomerase which mimics the substrate and binds covalently to the active site and then modifying the enzyme so it becomes irreversibly inhibited.

- Suicide inhibitors (Mechanism-based inhibitors) bind to the enzyme as a substrate and is processed by a normal catalytic mechanism that generates a chemically reactive intermediate that inactivates the enzyme through covalent modification. An example of a mechanism-based inhibitor can be seen through the inhibitory power of the planar proline intermediate formed during proline racemization. During this process, a trigonal intermediate is formed, and the formation of the racemase is inhibited because the tetrahedral intermediate necessary for formation of the product is not formed. The isomerization of proline through the planar transition state underscores the essence of transition-state analogs as potent inhibitors of enzymes.

Irreversible inhibition is covalent modification of enzymes such that the chemical reaction is not reversible; the inhibition molecules has specificity for their own enzyme to inactivate them, such that they work by changing active site of enzymes; the binding of enzyme to inhibitor forms enzyme complex that is reversible and not covalent that reacts to form another complex that can not work for catalysis reaction. The inhibition reaction can be changed by reversible competition of enzyme and substrate or other reversible inhibitor.

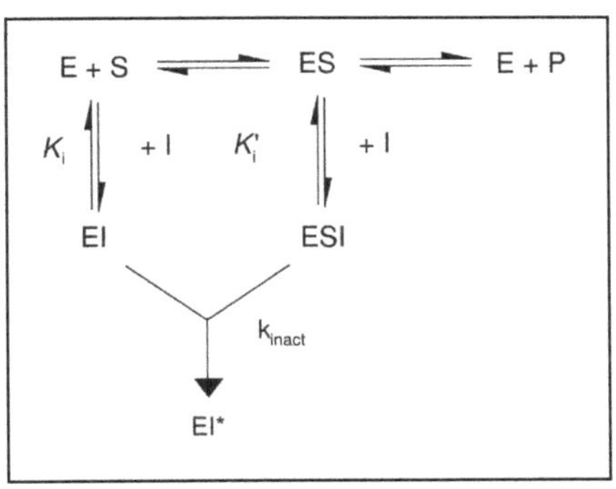

Kinetic scheme for irreversible inhibitors.

REVERSIBLE INHIBITORS

A reversible inhibitor inactivates an enzyme through noncovalent, more easily reversed, interactions. Unlike an irreversible inhibitor, a reversible inhibitor can dissociate from the enzyme. Reversible inhibitors include competitive inhibitors and noncompetitive inhibitors. A competitive inhibitor is any compound that bears a structural resemblance to a particular substrate and thus competes with that substrate for binding at the active site of an enzyme. The inhibitor is not acted on by the enzyme but does prevent the substrate from approaching the active site.

The degree to which a competitive inhibitor interferes with an enzyme's activity depends on the relative concentrations of the substrate and the inhibitor. If the inhibitor is present in relatively large quantities, it will initially block most of the active sites. But because the binding is reversible, some substrate molecules will eventually bind to the active site and be converted to product. Increasing the substrate concentration promotes displacement of the inhibitor from the active site. Competitive inhibition can be completely reversed by adding substrate so that it reaches a much higher concentration than that of the inhibitor.

Studies of competitive inhibition have provided helpful information about certain enzyme-substrate complexes and the interactions of specific groups at the active sites. As a result, pharmaceutical companies have synthesized drugs that competitively inhibit metabolic processes in bacteria and certain cancer cells. Many drugs are competitive inhibitors of specific enzymes.

A classic example of competitive inhibition is the effect of malonate on the enzyme activity of succinate dehydrogenase. Malonate and succinate are the anions of dicarboxylic acids and contain three and four carbon atoms, respectively. The malonate molecule binds to the active site because the spacing of its carboxyl groups is not greatly different from that of succinate. However, no catalytic reaction occurs because malonate does not have a CH2CH2 group to convert to CH=CH.

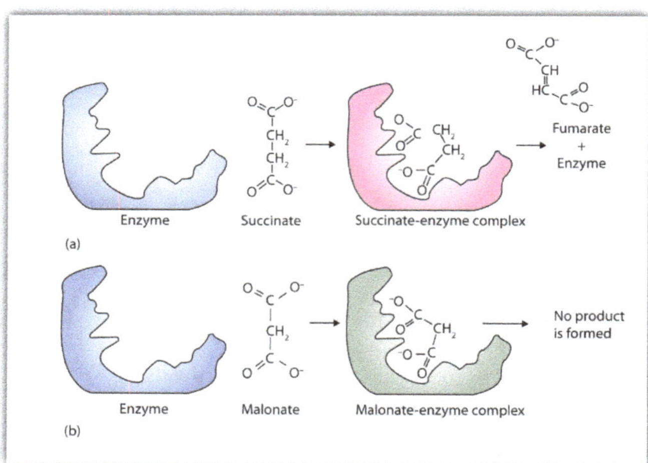

Competitive inhibitation.

Succinate binds to the enzyme succinate dehydrogenase. A dehydrogenation reaction occurs, and the product—fumarate—is released from the enzyme. Malonate also binds to the active site of succinate dehydrogenase. In this case, however, no subsequent reaction occurs while malonate remains bound to the enzyme.

Some reversible inhibitors are noncompetitive. A noncompetitive inhibitor can combine with either the free enzyme or the enzyme-substrate complex because its binding site on the enzyme is distinct from the active site. Binding of this kind of inhibitor alters the three-dimensional conformation of the enzyme, changing the configuration of the active site with one of two results. Either the enzyme-substrate complex does not form at its normal rate, or, once formed, it does not yield products at the normal rate. Because the inhibitor does not structurally resemble the substrate, the addition of excess substrate does *not* reverse the inhibitory effect.

References

- Mathews CK, van Holde KE, Appling DR, Anthony-Cahill SJ (February 26, 2012). Biochemistry (4 ed.). Pearson. ISBN 978-0138004644
- A-Enzyme-Inhibition, A-Enzyme-Kinetics, A-Physical-Chemistry-for-the-Biosciences, Physical-and-Theoretical-Chemistry: chem.libretexts.org, Retrieved 14 June, 2019
- Saha, Mrinmoy (2017). "Probing Mercaptobenzamides as HIV Inactivators via Nucleocapsid Protein7". ChemMedChem. 12 (10): 714–721. doi:10.1002/cmdc.201700141. PMC 5572807. PMID 28395128
- Enzyme-inhibition, introduction-to-chemistry-general-organic-and-biological: lardbucket.org, Retrieved 10 March, 2019
- Potashman MH, Duggan ME (March 2009). "Covalent modifiers: an orthogonal approach to drug design". Journal of Medicinal Chemistry. 52 (5): 1231–46. doi:10.1021/jm8008597. PMID 19203292

Applications of Enzymes

5 CHAPTER

Enzymes find applications in a variety of industries as extremely specific catalysts. Some of the industries where they are used are biofuel industry, dairy industry, brewing industry and paper industry. These diverse applications of enzymes have been thoroughly discussed in this chapter.

BIOFUEL INDUSTRY

Turning biomass into biofuel requires the breakdown of primary energy-rich component of the biomass, cellulose. Cellulose is a kind of polysaccharide, a compound formed from the bonding of many sugars into one. For the production of biofuels, the bonds between these many sugars must be broken so the energy could be generated by further processing these sugars. But breaking the bonds is not as easy as it sounds as the bonds are extremely strong and need a proper chemical reaction to complete the feat. Now the question arises that what are the candidates suitable for this job? The answer is various catalysts such as enzymes.

Use of Enzymes in Biofuel Production

Many enzymes are commonly used for the conversion of biomass into biofuels.

The nonrenewable sources of energy are speedily depleting from the face of earth and there is an inevitable demand for biofuels that can be developed from the abundant biomass currently available and that will continue to inhibit earth in abundance during any time in the future.

Other than the practically never ending feedstock for biomass on the face of earth and their renewable nature, there are many other environment related benefits attached with the concept of biofuels, such as less production of CO_2, reduced levels of carbon oxides, sulphur oxides and particulates and other emissions.

To match the production of the biofuels with their ever increasing demands, scientists are developing advanced biochemical processes by using enzymes. The research in this field is targeted at improving efficiency and reducing the negative impacts of the production processes on the environment as well as improving the quality of the biofuels produced.

Benefits of Enzymes

Enzymes are known to overcome many drawbacks associated with the use of traditional chemicals as catalysts for biofuel generation. For instance, the use of enzymes allows consumption of less energy in the production process. Enzymes can also utilize unrefined feedstock such as waste oil, without even needing the separation of free fatty acids that may be present in abundance in the feedstock, which is not possible with conventional catalysts. Also, if enzymes are used in the production of bioethanol from hydrolysis of cellulose, the utility cost of the process is greatly reduced as compared to conventional catalysts as no subsequent treatment of the byproduct is required and the process is carried out at mild conditions.

DAIRY INDUSTRY

Enzymes represent biological molecules and are proteinaceous in nature except catalytic ribozymes. They are used as activators in the biological system in order to enhance nearly all of the chemical reactions required to sustain and support life of organisms. Because of their easy availability and quick growth rate, microorganisms are considered preferable sources for industrial enzymes. Recently, different industries such as food, chemicals, and dairy are increasing their use of microbial enzymes in order to decrease processing time, reduce the required energy, and achieve cost effectiveness. The most important enzymes used in dairy products are lipases, lactoperoxidase, proteases, transglutaminase, lactase, esterases, catalase, lysozyme, aminopeptidase, etc. In addition, the enzymes of the dairy industry may be protective or coagulants that help in improving both the safety as well as the shelf life of different dairy products. The most known dairy industries use dairy enzymes in yogurt, cheese, butter, and other dairy products.

FOOD PROCESSING

The use of enzymes or microorganisms in food preparations is an age-old process. With the advancement of technology, novel enzymes with wide range of applications and specificity have been developed and new application areas are still being explored. Microorganisms such as bacteria, yeast and fungi and their enzymes are widely used in several food preparations for improving the taste and texture and they offer huge economic benefits to industries. Microbial enzymes are the preferred source to plants or animals due to several advantages such as easy, cost-effective and consistent production.

Microorganisms have been used in food fermentation since ancient times and fermentation processes are still applied in the preparation of many of the food items. Microbial enzymes play a major role in food industries because they are more stable than plant and animal enzymes. They can be produced through fermentation techniques in a cost-effective manner with less time and space requirement, and because of their high consistency, process modification and optimization can be done very easily. Many of these enzymes find numerous applications in various industrial sectors, e.g. amylolytic enzymes find applications in food, detergent, paper and textile industries. They are used for the production of glucose syrups, crystalline glucose, high fructose corn syrups, maltose syrups, etc. In detergent industry, they are used as additives to remove starch-based stains.

A-Amylases

α-Amylases are starch-degrading enzymes capable of hydrolyzing α-1,4 glycosidic bonds of polysaccharides, which results in the production of short-chain dextrins. These enzymes are widely distributed in all living organisms. Majority of α-amylases are metalloenzymes and require calcium ions for their activity, stability as well as integrity.

Wide applications of α-amylases in food industry include baking, brewing, starch liquefaction as well as a digestive aid. They are widely used in baking industry as flavour enhancement and antistaling agent to improve bread quality. During baking, α-amylases are added to the dough for conversion of starch to smaller dextrins, which are subsequently fermented by yeast. It improves the taste, crust colour and toasting qualities of bread.

α-Amylases are also used in the manufacture of high-molecular-mass branched dextrins. They are used as a glazing agent for the production of rice cakes and powdery foods. In starch industry, they also find application for starch liquefaction, which converts starch into glucose and fructose syrups. Enzymatic conversion of starch involves three steps: gelatinization, liquefaction and saccharification. Gelatinization involves formation of a viscous suspension by dissolution of starch granules. This is followed by a liquefaction process, which reduces viscosity and involves partial hydrolysis. Glucose and maltose are further produced by saccharification. This requires highly thermostable enzymes and most of the starch saccharification is carried out with α-amylases from Bacillus amyloliquefaciencs, Bacillus stearothermophilus or Bacillus licheniformis.

For the production of ethanol, starch is converted to fermentable sugars by the action of α-amylases and further fermentation of the sugars to alcohol is carried out by Saccharomyces cerevisiae. Other applications of α-amylases include clarification of fruit juices, which is carried out in the presence of cellulases and pectinases to improve yield as well as to make the process cost-effective.

Glucoamylases

Glucoamylases are exo-acting enzymes which catalyze the hydrolysis of polysaccharide starch from the non-reducing end, releasing β-glucose. They are also called saccharifying enzymes and are widely distributed in all living organisms. These enzymes are produced mainly by Aspergillus niger and Aspergillus awamori, but the one produced by Rhizopus oryzae is widely used for industrial applications. Majority of glucoamylases are stable at low temperature. At higher temperatures, they lose activity due to conformational change. Glucoamylases find wide range of applications in food industry, such as for the production of high-glucose syrups and high-fructose syrups. They also find application in baking industry to improve flour quality, reduce dough staling, as well as to improve bread crust colour and the quality of high fibre baked products. Glucoamylases convert the starch present in the flour to maltose and fermentable sugars. Fermentation by yeast leads to dough rise. These enzymes are also used for the production of glucose, which upon fermentation with Saccharomyces cerevisiae yields ethanol. Glucoamylases play an important role in the production of sake and soya sauce, as well as in the production of light beer. They metabolize dextrins and convert them to fermentable sugars with reduced calorific value and alcohol content in the beer.

Proteases

Proteases are enzymes which catalyze the hydrolysis of peptide bonds present in proteins and polypeptides. They are widely used in detergent and pharmaceutical, followed by food industries. They represent 60% of industrial enzymes on the market. The global demand for protease enzyme market has been growing at a compound annual growth rate (CAGR) of 5.3% during the period 2014-2019. Their demand is expected to increase much further as they find applications in leather processing as well as bioremediation processes. Proteases can be classified based on their origin, catalytic activity and nature of the reactive group in the catalytic site. The major sources of protease enzymes are animals, plant and microorganisms (both bacterial and fungal). Proteases are divided into two groups: exopeptidases and endopeptidases, based on the site of action on polypeptide chains. The exopeptidases act on the ends of polypeptide chains and endopeptidases act randomly in the inner regions of polypeptide chains. The endopeptidases are further classified into six groups, based on the catalytic residue present in the active site: Serine, aspartic, cysteine, metallo, glutamic acid and threonine protease.

Plant proteases such as bromelain, ficin and papain are widely used in food industry for various applications such as brewing, tenderization of meat, coagulation of milk and as a digestive aid. In addition, proteases are also used to improve the flavour, nutritional value, solubility and digestibility of food proteins as well as to modify their functional properties including coagulation and emulsification. Proteases are widely used in baking industry for the production of bread, baked foods, crackers and waffles. These enzymes are used to reduce the mixing time, decrease dough consistency and uniformity, regulate the gluten strength in bread and to improve the texture and flavour. The acid protease from Aspergillus usamii has been successfully employed for the improvement of functional properties of wheat gluten. The addition of protease could release sufficient peptides and amino acid levels in the wort to get a proper fermentation.

Acidic fungal proteases are used in improving fermentation of beer as they are efficient even at low pH by balancing the amino acid profile of beer. Another major application of proteases is associated with dairy industry. Naturally occurring proteases contribute significantly to the flavour characteristics of cheese. They are used for the acceleration of cheese ripening, to modify the functional properties and reduce the allergenic properties of milk products. In cheese making, proteases are also used to hydrolyze the specific peptide bond to generate paracasein and macropeptides.

Lactase (B-Galactosidase)

Hydrolysis of lactose is an important biotechnological process in food industry. The enzyme β-galactosidase catalyzes the hydrolysis of lactose. It belongs to the family of hydrolases. β-Galactosidase can be obtained from numerous biological systems including plants, animals and microorganisms. The production of β-galactosidase from microorganisms such as bacteria, fungi and yeast is a preferred choice due to higher yield and thus relatively low cost of the enzyme. The choice of source depends on the final application of the enzyme β-galactosidase, e.g. β-galactosidase from yeasts with pH optima of 6.5-7.0 is generally used for the hydrolysis of lactose in milk of whey. In the case of acidic whey hydrolysis, fungal β-galactosidase with pH optima of 3.0-5.0 is suitable. Thus the selection of β-galactosidase depends on the final application of the enzyme or industry. β-Galactosidase produced from yeast Kluyveromyces lactis requires ions such as Mn^{2+} or Na^{+}, where as Kluyveromyces fragilis requires Mn^{2+}, Mg^{2+}or K^{+}.

In industrial applications, two major classes of β-galactosidase are of prime importance. They are cold-active and thermostable β-galactosidase. On commercial scale, β-galactosidase is produced using microorganisms with GRAS status for their application in milk and dairy products. Lactase is used with milk and milk-based products to reduce lactose intolerance in people. The scoopability and creaminess of ice creams improved significantly after the hydrolysis of lactose with lactase. Additional advantage of hydrolyzing lactose into monomers is the reduction requirement of sweeteners as they could improve the sweetness of the products. Another major application of lactase is the lactose hydrolysis in whey. Whey is a byproduct of cheese production and its main components are lactose, proteins and minerals. This causes critical environmental issues associated with dairy industry as lactose is associated with high biological oxygen demand (BOD) and chemical oxygen demand (COD). Another application of lactase is the formation of galactooligosaccharides (GOS) from lactose hydrolysis due to transglycosylation activity of β-galactosidase. The GOS could be used as prebiotic food ingredients.

Lipases

Lipases are enzymes which catalyze the hydrolysis of long-chain triglycerides. They are naturally present in the stomach and pancreas of humans and other animal species in order to digest fats and lipids. Microbial lipases are produced by bacteria, fungi and yeast. Microbial enzymes contribute to approx. 90% of global lipase market. This enzyme finds application in various industries including food, biofuel, detergents and animal feed. In the food and beverage industry, lipases find major application in dairy, baking, fruit juice, beer and wine industries. Although it finds many applications in various industries, the market share of lipase is less than 10% of global industrial enzyme market.

Commercial lipases are mainly used for flavour development in dairy products and processing of other foods containing fat. They can improve the characteristic flavour of cheese by acting on the milk fats to produce free fatty acids after hydrolysis. Different types of cheese can be made by using lipases from various sources, e.g. Romano cheese using kid/lamb pre-gastric lipase, Camembert cheese using lipase from Penicillium camemberti and cheddar cheese using Aspergillus niger or A. oryzae. Lipase catalysis could improve the texture and softness of cheese. Lipases are also used as flavour development agents in butter and margarine, also to prolong the shelf life of various baking products. In alcoholic beverages such as wine, the aroma can be modified using lipase. They are used to improve the quality of cocoa butter, which has a melting point of 37 °C due to the presence of palmitic and stearic acids and can easily melt at 37 °C. A patent has been filed by Unilever using immobilized Rhizopus miehei lipase, which can replace palmitic with stearic acid to give desired stearic-oleic-stearic triglyceride. Functionalized phenols were esterified for the synthesis of lipophilic antioxidants for the application in sunflower oil using immobilized lipase from Candida antarctica (CALB), Candida cylindracea Ay30, Helvina lanuginosa, Pseudomonas sp. and Geotrichum candidum. Lipases also find application as a biosensor in food industry. Immobilized lipase was successfully used for the determination of organophosphorous pesticides with a surface acoustic wave impedance sensor by lipase hydrolysis. It may also be used in the determination of triglycerides and blood cholesterol if the lipase is immobilized onto pH/oxygen electrodes in combination with glucose oxidase. Microbial lipases such as lipase from Candida rugosa have many applications which cannot be met by chemical synthesis. This lipase finds application in the production of ice cream, single-cell protein, carbohydrate esters and amino acid derivatives. In addition to this, lipase could also be used in the processing of different waste streams that are released from food industries.

Phospholipases

Phospholipases selectively break down phospholipids into fatty acids and other lipophilic substances. They can be divided into four major classes (A, B, C and D) based on their mechanism of action. Phospholipase A1 (PLA1), phospholipase A2 (PLA2), and phospholipase B act on the carboxylic ester bonds of phospholipids, thus displacing and replacing the acyl group chain through various chemical reactions like hydrolysis, esterification and transesterification. Phospholipases C (PLC) and D (PLD), which modify polar head group, are also known as phosphodiesterases, and they recognize the phosphodiester linkage. Phospholipases are widely used in food industry, most importantly in the production of oils, dairy industry and in the manufacture of several bakery items. They also find applications in the degumming of various vegetable oils, cheese manufacture and bread manufacture. Phospholipase from Fusarium oxysporum is a commercially available phospholipase which has both phospholipase and lipase activities and it is marketed by Novozymes A/S (Denmark) for baking application under the name LipopanF®. The PLA2 commercialized by DSM Food Specialties (The Netherlands) with the trade name Maxapal® A2 has been reported by Zhao et al., who described that the egg yolk treatment with PLA could increase the stability of dough and interaction of starch and gluten. Another commercial phospholipase in baking industry is LysoMax® product (DSM Food Specialties), which is made up of a bacterial strain, specifically acting on lecithin. Phospholipases are also used in the processing of various dairy products to enhance the stability of fat or maximise the yield of cheese, milk, butter and ice cream. The important applications of lipases include enhancing the cheese flavour, lipolyzed milk fat production for use in butter as flavour, etc.

Esterases

In aqueous solution, esterases are able to facilitate the splitting of esters into acid and alcohol. In addition to this, esterases hydrolyze short-chain rather than long-chain acylglycerols, thus being different from lipases. Esterases play a prominent role in the food industry and alcoholic beverage industries, where they have been mostly used for the modification of oil and fat in various fruit juices and to produce fragrances and flavours. Feruloyl esterases, an important group of enzymes from esterase family, break the ester bond between ferulic acid and different polysaccharides in plant cell wall. Since feruloyl esterases hydrolyse lignocellulosic biomass, they are inevitable for waste management.

Cheng et al. screened for feruloyl esterase activity in a metagenomic library obtained from the microbial population of a cow rumen and identified a protease-resistant feruloyl esterase, which can release ferulic acid from wheat straw. This particular esterase has great commercial application because of its high pH and thermal stability and protease resistance.

In cheese manufacture, the fruity flavours are the result of different methyl or ethyl esters of short-chain fatty acids. Bacterial production of ethyl esters and thioesters has been reported. Alvarez-Macarie and Baratti reported the production of a novel thermostable esterase from the highly thermotolerant Bacillus licheniformis heterologously expressed in E. coli for the production of short-chain flavour esters. Feruloyl esterase is a key enzyme in the biosynthesis of ferulic acid, which is the precursor for vanillin, an aroma compound used in foods and beverages. Several researchers have reported the microbial production of ferulyl esterase.

Lipoxygenases

Lipoxygenases (LOX) are involved in the dioxygenation of polyunsaturated fatty acids in lipids containing a cis-1,4-pentadiene. They contain single polypeptide chain which is further assembled into an N-terminal domain and the catalytic β-barrel domain. LOX enzymes are non-haem iron-containing enzymes. The LOX-catalyzed reaction produces different precursors for the production of different volatile and aroma-producing chemical substances in plants. LOXs are used in aroma generation in food industry and also in bread making. Soya bean lipoxygenases (LOX) is the most studied lipoxygenase enzyme. Bacterial LOXs possess different specificity towards fatty acids. LOX from Nostoc sp. oxygenates at specific site in linoleic acid, but the LOX from Anabaena sp. exhibits variable specificity.

The main applications of LOXs in dough are based on their ability to bleach the flour pigment carotenoid, by co-oxidation of the pigment with fatty acids. Lipoxygenases are also employed to improve tolerance to mixing and different handling properties of dough. This effect is due to the oxidation of thiol group in gluten, which may lead to redistribution of different disulphide bonds, tyrosine cross-linking and subsequent strengthening of the gluten. This also leads to the improvement in dough rheology. Recently Patel et al. purified lipoxygenase from Lasiodiplodia theobromae by different chromatography techniques and fully characterized the enzyme. *L. theobromae* was reported to contain two types of lipoxygenases with molecular mass of 93 and 45 kDa.

Cellulases

Cellulases are enzymes that act on polymeric cellulose and hydrolyze β-1,4 linkages to liberate glucose units. The three major classes of cellulases are endo-(1,4)-β-d-glucanase, exo-(1,4)-β-d-glucanase and β-glucosidases. The catalytic modules of cellulases belong to glycosyde hydrolase (GH) family and have been classified in different groups based on differences in amino acid sequences and three-dimensional structural features. GH family enzymes mainly use acid–base catalysis mechanism for cleaving glycoside bonds in cellulose.

The catalysis is achieved by two major residues (a proton donor and a nucleophile) of the enzyme in the active site region. The hydrolysis occurs via retention or inversion mechanism depending on the spatial position of these catalytic residues in the enzyme. Endoglucanases cleave β-1,4-bonds in amorphous region of cellulose and expose the non-reducing and reducing ends of cellulosic polymer. Endoglucanases from various sources belong to different glycoside hydrolase families, among which the major are 5–9, 12, 44, 45, 48, 51 and 74. Most of the fungal endoglucanases contain a catalytic module with carbohydrate-binding module (CBM), but catalytic module without CBM was also reported from fungal species. Multiple catalytic modules and CBMs are present in bacterial endoglucanases. A cleft/grove-shaped active site is present in the catalytic module of most of the endoglucanases. Exoglucanases or cellobiohydrolases (CBHs) act on available reducing or non-reducing ends of cellulose polymer and liberate cellobiose. Fungal and bacterial CBHs show diversity in catalytic module and belong to glycoside hydrolase families 5, 6, 7, 9, 48 and 74.

A tunnel-shaped catalytic module is observed in most of the CBHs. β-Glucosidases catalyze the final step in cellulose breakdown by cleaving the non-reducing terminal β-d-glucosyl residues and removing β-d-glucose. The catalytic modules belonging to glycoside hydrolase families 1, 3 and 9 are reported from various β-glucosidases. The cellulolytic machinery of microbes is mainly regulated through feedback inhibition of b-glucosidases by their reaction product glucose. The major difference from CBHs is the absence of CBM in their structure. A pocket-shaped active site region of β-glucosidases helps them to attach the glucose molecule to non-reducing end and release glucose unit from cellodextrins or cellobioses. A large diversity of microorganisms is reported to produce cellulases during their growth on cellulosic materials. The industrial making of cellulases is mainly from microbial sources, bacteria and fungi, and these microorganisms can be diverse in their habitat.

The aerobic bacteria show similar mechanism of cellulose degradation to that of aerobic fungi. In anaerobic bacteria, cellulosomes are located on the cell surface and operate via a different system. Cellulases from fungi (Aspergillus and Trichoderma) and bacteria (Bacillus and Paenibacillus) are potentially used in the production of food. They are also widely used for various industries such as textile, paper, detergent and food industry. In juice industry, cellulases are applied in combination with other macerating enzymes for increasing process performance and yield, improving the extraction methods, clarification and stabilization of juices. They can also reduce the viscosity of nectar and puree from fruits such as apricot, mango, plum, papaya, pear and peach, and are used for the extraction of flavonoids from flowers and seeds. The preferences of cellulase-mediated extraction over conventional methods are due to higher yield, less heat damage and short processing time. Cellulases are utilized for the extraction of phenolic compounds from grape pomace. β-Glucosidases in combination with pectinase alter the structure, flavour and aroma of fruits and vegetables. They are also reported to reduce bitterness of citrus fruit and improve aroma and taste. Cellulases are used with other enzymes for efficient olive oil extraction. In wine production, cellulases are used in combination with other enzymes to increase yield and quality. The main advantages of using these enzymes are improved maceration, better colour development, must clarification and finally wine stability and quality. Studies of Oksanen et al. showed that cellulases can significantly reduce wort viscosity. The aroma of wines can be improved by β-glucosidases through modifications of glycosylated precursors.

Xylanases

Xylanases are produced by microorganisms to cleave xylans, a major constituent of hemicellulose. Three major enzymes, endoxylanases, exoxylanases and β-xylosidases, act synergistically and are required for the breakdown of xylan backbone in hemicellulose. Endoxylanases cleave the β-1,4 bonds of xylan backbone. Exoxylanases hydrolyse β-1,4 bonds of xylan from the non-reducing ends and release xylooligosaccharides. β-Xylosidases cleave the xylobiose and xylooligosaccharides to release xylose. The major functions of xylanases are performed by a catalytic module and few classes possess an additional CBM for binding to substrates. The two major catalytic modules of hemicellulases are glycoside hydrolases (GHs) and carbohydrate esterases (CEs). Endoxylanase hydrolyses the xylan backbone and has catalytic cores belonging to GH families 8, 10, 11, 30 and 43 with the most common ones being GH 10 and 11. These differ in their substrate specificities and the GH10 is more active on substituted xylan. Similar to cellulases, they may also contain CBMs. Exoxylanases randomly cleave the xylan backbone from inside, releasing long chain xylooligomers on which the β-xylosidases act. The catalytic module of these enzymes belongs to the GH families 3, 30, 39, 43, 52 and 54. These two enzymes are often collectively called xylanases. β-Xylosidase or xylan-1,4-β-xylosidase act on the xylooligosaccharides and xylobiose to release xyloses.

Xylanases are produced by microbes like actinomycetes, bacteria and fungi. The major actinomycete and bacterial species producing xylanase are Streptomyces sp., Bacillus sp. and Pseudomonas sp. Those produced by bacteria and actinomycetes are effective in a broader range of pH (5.0–9.0), with the optimum temperature for xylanase activity between 35 and 60 °C. Fungi are major sources of xylanase due to their high content and extracellular release of the enzyme. The major fungal species producing xylanase are Aspergillus sp., Fusarium sp. and Penicillium sp. Fungal xylanases have higher activity than bacteria or yeast.

Carbohydrate-hydrolyzing enzymes are usually used in bread making industry. Rheological properties of dough are improved through enzymatic hydrolysis of non-starch polysaccharides. Xylanases are widely used in bread making industry with other enzymes. The potential effectiveness of xylanolytic enzymes increases its use in bread making. They can increase the specific bread volume and this improves the quality of bread. The hemicellulose in wheat flour is broken down by xylanase, which increases the binding of water in the dough. The dough becomes softer and crumb formation is delayed, allowing the dough to grow. Xylanase is used to improve texture, tastiness and palatability in biscuits. They also play an important role in juice production by improving extraction, clarification and stabilization. In combination with other enzymes, xylanases lead to better yield of juice and increased recovery of aromas, essential oils, vitamins, mineral salts, pigments, etc. In beer making industries, xylanases are used for hydrolysing the cellular wall of barley. Hydrolysis leads to release of arabinoxylans and lower oligosaccharides, which reduces the muddy appearance and viscosity of the beer.

Pectinases

Pectinases are enzymes which catalyze the hydrolysis of glycosidic bonds in pectic polymers. Pectic substances found in tomato, pineapple, orange, apple, lemon pulp, orange peel and other citrus fruits act as natural substrate for this enzyme. Functionally pectinases can be categorized as polygalacturonases (which hydrolyse glycosidic α-(1-4) bonds), pectin esterases (which remove acetyl and methoxyl groups from pectin), pectin lyase and pectate lyase. Pectinases can be produced from natural as well as recombinant microbes with attempts made to increase their thermostability and yield. Pectinases can also act either on smooth or hairy regions of pectin. Based on pH, there are acidic and alkaline pectinases also grouped in endopectinases when enzyme cleaves randomly, and exopectinases when the terminal ends are targeted.

Pectinases find a multitude of industrial applications such as in, food industry, remediation, etc. Juices with added pectinase have a clearer appearance and filterability than enzyme-depleted counterparts. Apart from reducing the turbidity and haze generation of naturally derived fruit juices such as apple and banana, pectinases also improve the colour and flavour of drinks. The addition of gelatin and pectin greatly increases the viscosity and turbidity of juices, and removal of the haze is the most costly part of juice production. The use of biogenic enzymes such as pectinases in juices would act almost nine times better than mechanical maceration to get good results.

Glucose Oxidase

Glucose oxidase belongs to a large group of enzyme family called oxidoreductases. Glucose oxidase is a flavoprotein discovered in 1928 by Müller. He stated that in the presence of dissolved oxygen the enzyme can convert glucose to gluconic acid. In the reaction, β-d-glucose is oxidised to gluconolactone and molecular oxygen is reduced to hydrogen peroxide. The glucanolactone is then spontaneously hydrolysed to gluconic acid. The enzyme is homodimeric and contains two similar polypeptide chain subunits (80 kDa). The subunits are covalently linked by disulphide bonds and one flavin adenine dinucleotide (FAD) molecule non-covalently bound to active site region of each subunit. The glucose oxidase production has been reported from various microorganisms and it was first discovered in Aspergillus niger and Penicillium glaucum. Aspergillus niger species is widely used for production of glucose oxidase and its strains can produce higher amount of glucose oxidase. Penicillium adametzii is a widely used fungus for the production of extracellular glucose oxidase. The different bacterial species are also reported to produce glucose oxidase. Although many species of bacteria and fungi are reported to produce this enzyme, fungi are considered for the industrial production of glucose oxidase.

Glucose oxidase has its wide use in various industries like pharmaceutical and food industries, and in biofuel cells. Its use is increasing in baking industry because its oxidizing effects make stronger dough. In food industry, it enhances the flavour, aroma and stability of food products by removing glucose and oxygen from diabetic drinks and egg white. Glucose oxidase improves the colour, texture, flavour and shelf life of food products and prevents rottening. During food packaging glucose oxidase is used for increasing storage life by removing oxygen.

Laccase

Laccases are a cluster of oxidases which represent the largest subgroup of multicopper enzymes. Commonly known as blue oxidases, they are used for studying their potential to oxidize phenolic compounds and therefore applied in several industrial sectors.

These enzymes act as a potent biocatalyst for application in chemical synthesis, biobleaching of paper pulp, bioremediation, biosensing, wine stabilization and textile finishing. They have different specificity for substrate and a wide range of oxidizable substrates, which further depends on the type of microbial sources producing the enzyme. Laccases catalyze the oxidation of a wide range of compounds such as phenolics, aromatic amines and ascorbate. These enzymes combine reducing substrate having four oxidized electrons with four reduced electrons for cleaving dioxygen bond in the presence of four copper atoms present in laccases.

Laccases are secreted extracellularly by several fungi as a product of their secondary metabolism during fermentation, but their production is limited to a few fungal species. Well known producers of laccases belong to Deuteromycetes, Ascomycetes and Basidiomycetes. Funalia trogii is a white rot fungus capable of producing laccase through absorbent fermentation. The maximum laccase production by F. trogii reached 11 900 U/L, which was 4.97 times higher than that of normal fermentation. Bacillus licheniformis produces recombinant laccases for industrial applications. Recently, heterologous expressions have been used for laccase production. Bacillus vallismortis fmb-103 genes were cloned and heterologously expressed in Escherichia coli BL21 (DE3) cells.

Laccase is used for modification of colour appearance of food and beverage industries, or for wine stabilization as an alternative to physical and chemical adsorbents. Removal of polyphenols from wine should be chosen to avoid adverse changes in wine organoleptic characteristics including stability in acidic medium and reversible inhibition due to the presence of sulphite. Furthermore, this enzyme is used in cork stopper manufacturing industry.

Haze formation is one of the problems in brewing industry. To avoid it, laccases have been applied for polyphenol oxidation as substitute for traditional approach by different researchers. This enzyme is also used for oxygen removal in the final step of beer production which prolongs the storage life of beer. Commercial laccase called Flavourstar, manufactured by Novozymes, is used for removing the off-flavour formation in brewing industry. It is used in baking because it has the capability to cross-link with biopolymers. The application of laccase in baking enhances stability, strength and decreases stickiness which further increases machinability of bread batter. Moreover, it increases volume and enhances softness of the product as reported by Labat et al. and Si.

Catalase

Catalase is a tetrameric protein found in aerobic organisms. It helps hydrogen peroxide decomposition. This enzyme can be produced from microbial sources such as Aspergillus niger and Micrococcus luteus and from bovine liver. Microorganisms are usually preferred as sources for enzyme production due to their advantages such as fast growth, easy handling and genetic tuning for obtaining a desired product. The anaerobic Bacteroides fragilis exhibited increased catalase levels in media with haem. Frankenberg et al. isolated catalase from Enterococcus faecalis, which completely depends on haem source without which it cannot synthesize porphyrin group. A facultative anaerobic catalase-producing Bacillus maroccanus resistant to hydrogen peroxides was isolated from textile effluents. A potent catalase-producing bacterium Pyrobaculum calidifontis was isolated from hot springs in Los Banos and Calamba, Laguna, Philippines. A thermo-alkaliphilic catalase-positive strain of Bacillus halodurans LBK 261 was isolated from alkaline hot-spring waters of Kenya. A halo (alkali) tolerant catalase-producing Oceanobacillus oncorhynchi ssp. incaldaniensis was isolated from an algal mat capable of producing catalase at wide range of pH 6.0–9.5 and salinity of 5–20%. A catalase-positive psychrophile Bacillus N2a was isolated from seawater. Other catalase-positive bacteria such as Rhizobium radiobacter were isolated from industrial effluent from beverage industry, Comamonas testosteroni and C. terrigena from effluent sludge enriched with crude oil along with heavy metals and Serratia SYBC08 from hydrogen peroxide sludge. Psychrobacter piscatorii T-3, a psychrotolerant bacterium isolated from bleach-rich runoff, has high catalase activity. Fungi and yeast are able to produce catalase. The highest level of catalase activity of 400 mg/g was observed in isolates of Aspergillus niger. A catalase-positive entomopathogenic fungus Metarhizium anisopliae strain Ma10 (CNRCB MaPL10) was isolated from Geraeus senilis.

In fabric industry, catalase is used for removing excess hydrogen peroxide from fabric. This enzyme is mostly used along with other enzymes in food processing industry. Catalase is often used with glucose oxidases for food preservation. Ough used a glucose oxidase/catalase cocktail for elimination of oxygen from wine before bottling and evaluated the formation of acetaldehydes. Results showed that colour and amount of acetaldehyde were stable if treated properly with enzymes. Catalase is applied in milk processing industry to eliminate peroxide from milk, to remove glucose from egg white in baking industry and in food wrappers to prevent oxidation and control perishability of food. This enzyme has limited use in cheese production.

Peroxidase

Peroxidases are oxidoreductase proteins that contain iron (III) protoporphyrin IX as the prosthetic group. They catalyse the reduction of peroxides and oxidation of a wide range of inorganic and organic compounds. Their molecular mass ranges from 30 000 to 150 000 Da, and they comprise a group of unique enzymes such as iodide peroxidase, NADH peroxidase and glutathione peroxidase as well as a group of other nonspecific enzymes. Peroxidases are present in plants, microorganisms and animals. They are involved in lignification processes in plants and defence mechanisms against damaged or infectious tissues.

Among microorganisms, Phanerochaete chrysosporium is the best characterized peroxidase-secreting organism. Industrial scale applications of fungal peroxidases are limited by challenges associated with post-translational modification of proteins. However, bacterial peroxidases are easier to produce and have better stability and activity suitable for industrial applications. These enzymes are applied with bacterial laccases for dye decolourization. Peroxidase activities are reported in bacterial taxa, such as Firmicutes, Proteobacteria, Actinobacteria and Acidobacteria. Moreover, actinomycetes, which are soil bacteria, are able to grow like fungi and have similar ecological niche, and can produce peroxidases for lignin degradation. The first secreted extracellular lignin peroxidase was produced by Streptomyces viridosporus T7A.

Peroxidase catalyzes a wide range of substrates using hydrogen peroxide or other peroxides. This enzyme is used in food industry for producing flavour, colour and texture and improving nutritional quality of food. Other applications include as biosensors, in polymer synthesis and in the management of pollutants in the environment. It can be used for treating phenolic effluents from industries. Thermal inactivation of peroxidases is used in food industry to measure the efficiency of blanching treatment, which further enhances the shelf life of food. The negative effect of peroxidases is that they cause undesirable browning of fruits and off-flavours of vegetables.

A-Acetolactate Decarboxylase

α-Acetolactate decarboxylase greatly aids in the fast maturation of beer. This enzyme can be produced from natural microbes such as Brevibacillus brevis or from recombinant Saccharomyces cerevisiae. The enzyme catalytically converts acetolactate to acetoin via a two-step reaction involving direct decarboxylation of substrate to an enol derivative and its further protonation to final product.

Enzyme-based removal of α-acetolactate and α-aceto-α-hydroxybutyrate assists in overcoming the rate-limiting step of beer maturation. While the maturation of beer without the use of enzymes takes 2 to 12 weeks, the use of α-acetolactate decarboxylase results in maturation within 24 hours depending on the source of enzyme. Moreover, the off-taste due to the presence of diacetyl in beer is nullified by the action of this enzyme. Studies indicate that both free and encapsulated form of this enzyme work efficiently in the process, thus aiding the use of immobilized enzymes at reduced costs. Novel inorganic nanoflowers or alginate microbeads immobilized with α-acetolactate decarboxylase are promising strategies with better thermal stability, reusability and catalytic efficiency.

Asparaginase

Of the various microbially derived enzymes, asparaginases form a major class of pharmaceutical, nutraceutical and industrially significant enzymes widely used by man. Asparaginase, as the name implies, catalyses the breakdown of the asparagine to subsequent acid derivative aspartic acid and NH_3 and can be considered as the asparagine-depleting enzyme. Asparagine is a nonessential amino acid to humans, whereas it is an essential amino acid for cancerous cells. Thus, the depletion of asparagine critically affects the growth of cancerous cells, which forms the basis of this enzyme as anticancer agent.

Various food processing methods such as frying in oil and baking cause the conversion of asparagine to acrylamide, a known carcinogen. Among various methods attempting to overcome the acrylamide formation, the depletion of asparagines by enzymatic treatment has been found effective in reducing the formation of acrylamides from asparagines by 97%.

Debittering Enzymes – Naringinase

Naringinase is mainly responsible for the breakdown of naringin, the principle bitter flavanone glycoside found in citrus fruits. Naringin is broken down to a glycon naringenin and rhamnose as a result of its α-rhamnosidase and β-glucosidase action. Nariginase is produced mostly by fungal isolates, viz. Aspergillus niger, Circinella, Eurotium, Fusarium, Penicillium, Rhizopus and Trichoderma and bacteria such as Bacillus sp., Burkholderia cenocepacia, Bacteriodes distasonis, Thermomicrobium roseum, Pseudomonas paucimobilis, etc. Fungal sources of naringinase are found to be more predominantly used than the bacterial ones due to increased yield.

Naringinase has a major role in food processing as a debittering enzyme supplemented to fruit juices. Both free and immobilized forms of this enzyme are used to get better results. Immobilization of this enzyme has been done in a variety of substrates such as polyvinyl alcohol cryogels, packaging films, cellulose triacetate nanofibre, graphene, etc. Various food additives such as biopolymers and sweeteners can be synthesized using rhamnosidase or naringinase. Yet, another use of naringinase together with β-glucosidase and arabinosidase is to improve the aroma of wine. The use of naringinase is also noted in tomato pulp preparation, kinnow peel waste treatment and prunin preparation.

BREWING INDUSTRY

Germinating barley kernel.

Beer and wine are both alcoholic beverages which are produced by yeast fermentation of sugars. Wine is based on grapes, and beer is traditionally based on barley. The matured grapes already contain the sugars needed for the fermentation, while barley contain starch that has to be broken down to fermentable sugars before the yeast can make alcohol. Therefore, traditional brewing contains and extra step compared with wine-making, namely malting in which enzymes needed for the degradation of starch into fermentable sugars are produced.

Malt is germinated barley or other cereals like wheat and sorghum: First the grains are "steeped" bringing the water content from about 12% to 45%, then they are allowed to germinate for 4-6 days and finally the germination is stopped by heating (kilning) reaching a final moisture content of about 4%. Some enzymes are already present in the barley, e.g. β-amylases, but the majority of enzymes are produced during the germination, e.g. α-amylases and proteases, and in the final malt all the enzymes needed for the conversion of "grains" into a fermentable liquid (wort) is present.

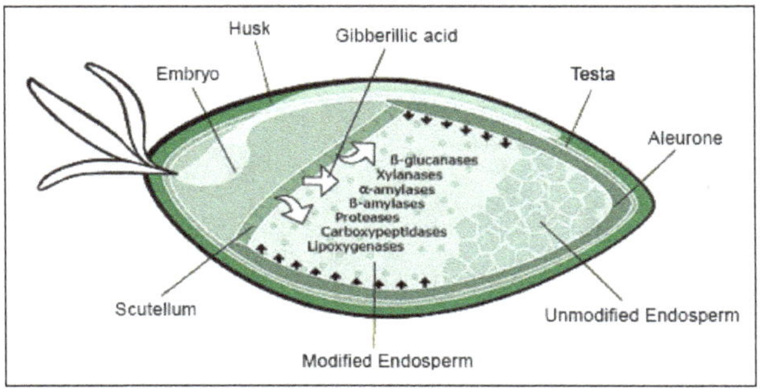

Enzyme production during malting.

In former days, production of malt was an integrated part of every brewery, but to day most malt is produced outside the brewery in large malt factories, and malt has become a purchased raw material, like other raw materials. This means that the breweries to day are more flexible in the use raw materials, and for that matter for the source of enzymes.

The malt enzymes do have some limitations. They can only work at certain temperatures, pH values etc., and the activities might be too low to do a proper job in proper time. In contrast, commercial exogenous enzymes can be designed to work at preferred temperatures and pH values, to have more enzymatic power, or to express wanted enzyme activities that are not present in malt. Addition of exogenous enzymes at various steps during the brewing process can therefore make brewing easier, faster and more consistent. It gives the brewmasters extra flexibility in the choice of raw materials due to less dependence on malt enzymes, as well as providing opportunity to create new products, which is not possible to make with malt enzymes alone. Also the possibility to improve beer quality by avoiding off-flavours is possible with commercial enzymes. The increasing concern on resources and CO_2- emission has also put the use of commercial enzymes within the brewing industry in focus. By the use of exogenous enzymes more can be extracted from the raw materials, more local raw materials can be used, and more unmalted grains can be used, saving significant amounts of energy and transport.

The Brewing Process

Traditionally, beer is produced by mixing crushed barley malt and hot water in a mash copper to perform the mashing. Besides malt, other starchy cereals such as maize, sorghum, rice and barley, or pure starch itself, can be added to the mash. These are known as adjuncts.

The standard mashing for pilsner type beer consists of several temperature steps, each favouring different malt enzyme activities. The lowest temperature (45 °C) is the optimal temperature for cell wall degrading enzymes, β-glucanases. The proteases works best at 52 °C, the β-amylase best at 63 °C and the α-amylase at 72 °C. The last step in the mashing is inactivation of the enzymes at 78 °C.

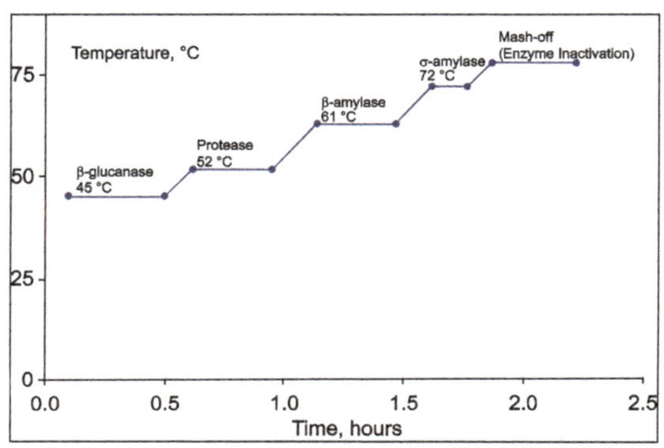

The traditional mashing temperature profile is determined by the temperature optima for the various malt enzymes.

If β-glucan and protein are properly broken down during malting, single temperature mashing at 65-71 °C has shown to be sufficient, as in the case of traditional ale brewing.

During mashing the starch is degraded to dextrin and fermentable sugars. α-amylase liquefy the gelatinized starch by hydrolysis of the α-1,4 linkages at random. β-amylases are exo-enzymes which attack the liquefied starch chains resulting in successive removal of maltose units from the non-reducing end.

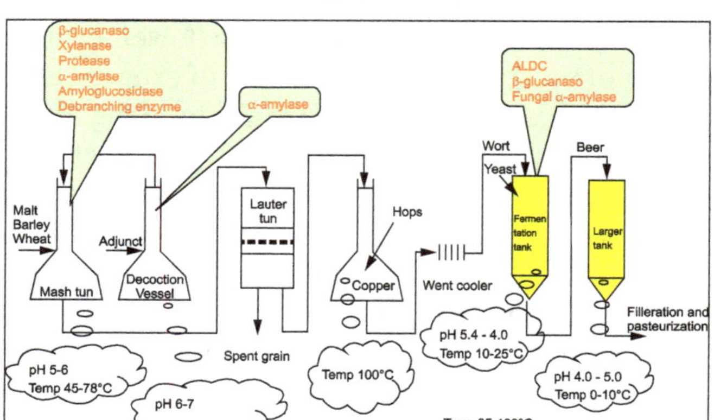

The processing steps in brewing where exogenous enzymes can be added.

After mashing, the mash is sieved in a lauter tun or on a mash filter. The resulting liquid, known as sweet wort, is then transferred to the copper, where it is boiled with hops. The hopped wort is cooled and transferred to the fermentation vessels, where yeast is added. In normal wort 2/3 of the carbohydrates are fermentable sugars. After fermentation, the so-called 'green beer' is matured before final filtration and bottling. Figure shows a diagram of the brewing process and where external enzymes are used for process aids.

Commercial Enzymes from Exogenous Sources

The traditional source of enzymes used for the conversion of cereals into beer is barley malt. If too little enzyme activity is present in the mash, there will be several undesirable consequences: the extract yield will be too low; wort separation will take too long; the fermentation process will be too slow; too little alcohol will be produced; the beer filtration rate will be reduced; and the flavour and stability of the beer will be inferior.

Exogenous enzymes are used to supplement the malt's own enzymes in order to prevent these problems. Furthermore, industrial enzymes are used to ensure better adjunct liquefaction, to produce low-carbohydrate beer ('light beer'), to shorten the beer maturation time, and to produce beer from cheaper raw materials.

The various steps of the brewing operations, where microbial enzymes are occasionally added, are shown in table.

Enzymes at Work

Quality and supply constraints on malt, and doubling of malt prices have given increased interest for enzyme solutions.

Exchanging part of the malt with barley has been popular because using barley in combination with commercial enzymes gives the same beer quality as with malt.

Introducing a higher content of starch hydrolysing enzymes offer the possibilities of producing "light beer" also called "low calorie beer".

An enzyme solution for diacetyl control after fermentation improves vessel utilization, save energy and ensures a high beer quality after a reduced maturation time.

Table: Steps of the brewing operations where microbial enzymes are used.

Operation	Enzymes	Enzyme action	Function
Decoction vessel (cereal cooker)	α-amylase	Hydrolyse starch.	Adjunct liquefaction Reduce viscosity.
	β-glucanase	Hydrolyse glucans.	Aid the filtration.

Mashing	α-amylase	Hydrolyse starch.	Malt improvement.
	Amyloglucosidase	Increase glucose content.	Increase % fermentable sugar in "light" beer.
	Debranching enzyme	Hydrolyse α-1,6 branch points of starch.	Secures maximum fermentability of the wort.
	Proteases	Increase soluble protein, and free amino- nitrogen (FAN).	Malt improvement Improved yeast growth.
	β-glucanase	Hydrolyse glucans.	Improve wort separation.
	Pentosanase/xylanase	Hydrolyse pentosans of malt, barley, wheat.	Improve extraction and beer filtration.
Fermentation	Fungal α-amylase	Increase maltose and glucose content.	Increase % fermentable sugar in "light" beer.
	β-glucanase	Hydrolyze glucans.	Reduce viscosity and aid filtration.
	α-acetolactate- decarboxylase (ALDC)	Converts α-acetolactate to acetoin directly.	Decrease fermentation time by avoiding formation of diacetyl.
Conditioning tank	Protease	Modify protein-polyphenolic compounds.	Reduce the chill haze formed in beer.

Brewing with Barley

Traditionally, the use of barley has been limited to 10-20% of the grist when using high-quality malts. At higher levels of barley or using undermodified malts, processing becomes more difficult. In these cases the mash needs to be supplemented with extra enzyme activity if the brewer is to benefit from the advantages of using unmalted barley while still maintaining brewing performance.

Brewers can either add a malt-equivalent blend of α-amylase, β-glucanase and protease at the mashing-in stage or add the enzymes separately as required.

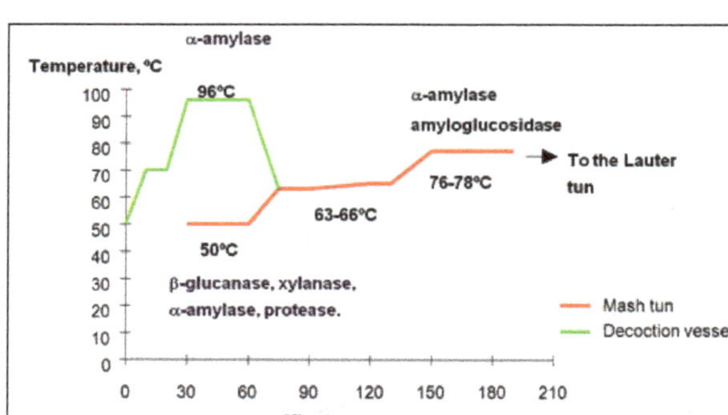

Mashing diagram for barley brewing.

The maize grits are liquefied separately with help of the α-amylase Termamyl BrewQ at 96 °C for 30 minutes, through a short holding time at 70 °C. It is stabilised by approximately 100 ppm Ca^{++} at a water-to-adjunct ratio of approximately 4:1. Milled malt and barley are mashed-in at a temperature of 50 °C. After 30 minutes the adjunct mash from the decoction vessel is added to increase the temperature to 63-66 °C. After 60 minutes the mash is heated to hold at 76-78 °C until starch-negative (no blue colour is formed with iodine in potassium iodide). Hereafter the wort separation is made in the lauter tun.

Enzymes to Improve Fermentation

Small adjustments in fermentability can be achieved by adding amyloglucosidase alone or in combination with debranching enzymes at mashing-in or a fungal α-amylase at the start of fermentation.

To describe to which extent the extracted sugars are fermentable brewers define degree of attenuation, which is synonymously with degree of fermentation or fermentability.

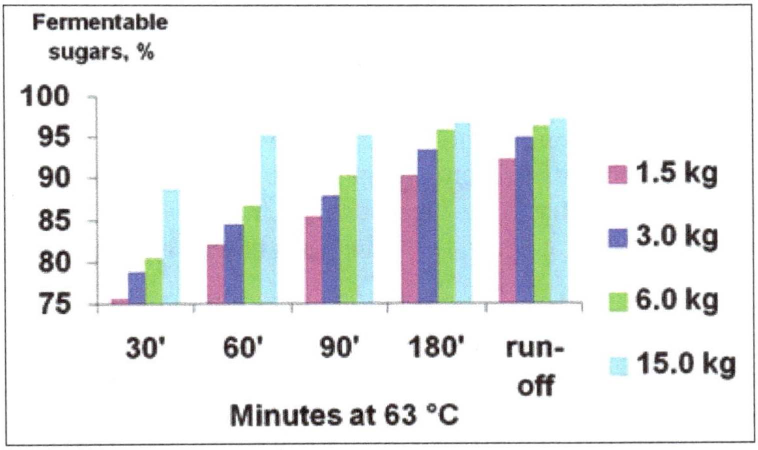

Total fermentable sugar production with different dosages of Attenuzyme® (kg per ton malt) and extended mashing at 63 °C.

Beer types with very high attenuation ("light beer" or "low calorie beer") are most often produced using amyloglucosidase alone. Extended mashing at 63 °C and high dosages of enzymes is necessary to produce extremely high attenuated beer.

Fungal α-amylases are used to produce mainly maltose and dextrins whereas amyloglucosidase produces glucose from both linear and branched dextrins.

Diacetyl Control

An important question for brewers is "When exactly is a beer mature?", because this determines when they can "rack" the beer to make way for the next batch. The simple answer to the above question is when the diacetyl level drops below a certain limit (about 0.07 ppm). Diacetyl gives beer an off-flavour like buttermilk and one of the main reasons for maturing a beer is to allow the diacetyl to drop to a level where it can't be tasted.

Diacetyl is formed by the non-enzymatic oxidative decarboxylation of α-acetolactate, which is produced by the yeast during primary fermentation. The yeast removes the diacetyl again during the beer maturation stage by conversion to acetoin, which has a much higher flavour threshold value. In fact, acetoin is almost tasteless compared with diacetyl.

By adding the enzyme α-acetolactate decarboxylase (ALDC) (e.g. Novozymes' Maturex) at the beginning of the primary fermentation process, it is possible to bypass the diacetyl step and convert α-acetolactate directly into acetoin. Most of the α-acetolactate is degraded before it has a chance to oxidise and less diacetyl is therefore formed. The brewery enjoys greater fermentation and maturation capacity without investing in new equipment.

BIOLOGICAL DETERGENT

The use of enzymes in detergent formulations is now common in developed countries, with over half of all detergents presently available containing enzymes. In spite of the fact that the detergent industry is the largest single market for enzymes at 25 - 30% of total sales. details of the enzymes used and the ways in which they are used, have rarely been published.

Dirt comes in many forms and includes proteins, starches and lipids. In addition, clothes that have been starched must be freed of the starch. Using detergents in water at high temperatures and with vigorous mixing, it is possible to remove most types of dirt but the cost of heating the water is high and lengthy mixing or beating will shorten the life of clothing and other materials. The use of enzymes allows lower temperatures to be employed and shorter periods of agitation are needed, often after a preliminary period of soaking. In general, enzyme detergents remove protein from clothes soiled with blood, milk, sweat, grass, etc. far more effectively than non-enzyme detergents. However, using modern bleaching and brightening agents, the difference between looking clean and being clean may be difficult to discern. At present only proteases and amylases are commonly used. Although a wide range of lipases is known, it is only very recently that lipases suitable for use in detergent preparations have been described.

Detergent enzymes must be cost-effective and safe to use. Early attempts to use proteases foundered because of producers and users developing hypersensitivity. This was combatted by developing dust-free granulates (about 0.5 mm in diameter) in which the enzyme is incorporated into an inner core, containing inorganic salts (e.g., NaCI) and sugars as preservative, bound with reinforcing, fibres of carboxymethyl cellulose or similar protective colloid. This core is coated with inert waxy materials made from paraffin oil or polyethylene glycol plus various hydrophilic binders, which later disperse in the wash. This combination of materials both prevents dust formation and protects the enzymes against damage by other detergent components during storage.

Enzymes are used in surprisingly small amounts in most detergent preparations, only 0.4 - 0.8% crude enzyme by weight (about 1% by cost). It follows that the ability to withstand the conditions of use is a more important criterion than extreme cheapness. Once released from its granulated form the enzyme must withstand anionic and non-ionic detergents, soaps, oxidants such as sodium perborate which generate hydrogen peroxide, optical brighteners and various less-reactive materials, all at pH values between 8.0 and 10.5. Although one effect of incorporating enzymes is that lower washing temperatures may be employed with consequent savings in energy consumption, the enzymes must retain activity up to 60 °C.

Table: Compositions of an enzyme detergent

Constituent	Composition (%)
Sodium tripolyphosphate (water softener, loosens dirt)	38.0
Sodium alkane sulphonate (surfactant)	25.0
Sodium perborate tetrahydrate (oxidising agent)	25.0
Soap (sodium alkane carboxylates)	3.0
Sodium sulphate (filler, water softener)	2.5
Sodium carboxymethyl cellulose (dirt-suspending agent)	1.6
Sodium metasilicate (binder, loosens dirt)	1.0
Bacillus protease (3% active)	0.8
Fluorescent brighteners	0.3
Foam-controlling agents	Trace
Perfume	Trace
Water	to 100%

The enzymes used are all produced using species of Bacillus, The a-amylase supplied for detergent use is Termamyl, the enzyme from B. licheniformis which is also used in the production of glucose syrups. a-Amylase is particularly useful in dish-washing and de-starching detergents.

In addition to the granulated forms intended for use in detergent powders, liquid preparations in solution in water and slurries of the enzyme in a non-ionic surfactant are available for formulating in liquid 'spotting' concentrates, used for removing stubborn stains. Preparations containing both Termamyl and Alcalase are produced, Termamyl being sufficiently resistant to proteolysis to retain activity for long enough to fulfil its function.

It should be noted that all the proteolytic enzymes described are fairly non-specific serine endoproteases, giving preferred cleavage on the carboxyl side of hydrophobic amino acid residues but capable of hydrolysing most peptide links. They convert their substrates into small, readily soluble fragments which can be removed easily from fabrics. Only serine protease; may be used in detergent formulations: thiol proteases (e.g., papain) would be oxidised by the bleaching agents, and metalloproteases (e.g. thermolysin) would lose their metal cofactors due to complexing with the water softening agents or hydroxyl ions.

The enzymes are supplied in forms suitable for formulation by detergent manufacturers. Domestic users are familiar with powdered preparations but liquid preparations for home use are increasingly available. Household laundering presents problems quite different from those of industrial laundering: the household wash consists of a great variety of fabrics soiled with a range of materials and the user requires convenience and effectiveness with less consideration of the cost. Home detergents will probably include both an amylase and a protease and a lengthy warm-water soaking time will be recommended. Industrial laundering requires effectiveness at minimum cost so heated water will be re-used if possible. Large laundries can separate their 'wash' into categories and thus minimise the usage of water and maximise the effectiveness of the detergents. Thus white cotton uniforms from an abattoir can be segregated for washing, only protease being required. A pre-wash soaking for 10-20 min at pH up to 11 and 30°-40 °C is followed by a main wash for 10-20 min at pH 11 and 60°-65 °C. The water from these stages is discarded to the sewer. A third wash includes hypochlorite as bleach which would inactivate the enzymes rapidly. The water from this stage is used again for the pre-wash but, by then, the hypochlorite concentration is insufficient to harm the enzyme. This is essentially a batch process: hospital laundries may employ continuous washing machines, which transfer less-initially-dirty linen from a pre-rinse initial stage, at 32 °C and pH 8.5, into the first wash at 60 °C and pH 11, then to a second wash, containing hydrogen peroxide, at 71 °C and pH 11, then to a bleaching stage and rinsing. Apart from the pre-soak stage, from which water is run to waste, the process operates counter-currently. Enzymes are used in the pre-wash and in the first wash, the levels of peroxide at this stage being insufficient to inactivate the enzymes.

There are opportunities to extend the use of enzymes in detergents both geographically and numerically. They have not found widespread use in developing countries which are often hot and dusty, making frequent washing of clothes necessary. The recent availability of a suitable lipase may increase the quantities of enzymes employed very significantly. There are, perhaps, opportunities for enzymes such as glucose oxidase, lipoxygenase and glycerol oxidase as means of generating hydrogen peroxide in situ. Added peroxidases may aid the bleaching efficacy of this peroxide.

A recent development in detergent enzymes has been the introduction of an alkaline-stable fungal cellulase preparation for use in washing cotton fabrics. During use, small fibres are raised from the surface of cotton thread, resulting in a change in the 'feel' of the fabric and, particularly, in the lowering of the brightness of colours. Treatment with cellulase removes the small fibres without apparently damaging the major fibres and restores the fabric to its 'as new' condition. The cellulase also aids the removal of soil particles from the wash by hydrolysing associated cellulose fibres.

PAPER INDUSTRY

The pulp and paper industry processes huge quantities of lignocellulosic biomass every year. The technology for pulp manufacture is highly diverse, and numerous opportunities exist for the application of microbial enzymes. Historically, enzymes have found some uses in the paper industry, but these have been mainly confined to areas such as modifications of raw starch. However, a wide range of applications in the pulp and paper industry have now been identified. The use of enzymes in the pulp and paper industry has grown rapidly since the mid 1980s. While many applications of enzymes in the pulp and paper industry are still in the research and development stage, several applications have found their way into the mills in an unprecedented short period of time. Currently the most important application of enzymes is in the prebleaching of kraft pulp. Xylanase enzymes have been found to be most effective for that purpose. Xylanase prebleaching technology is now in use at several mills worldwide. This technology has been successfully transferred to full industrial scale in just a few years. The enzymatic pitch control method using lipase was put into practice in a large-scale paper-making process as a routine operation in the early 1990s and was the first case in the world in which an enzyme was successfully applied in the actual paper-making process. Improvement of pulp drainage with enzymes is practiced routinely at mill scale. Enzymatic deinking has also been successfully applied during mill trials and can be expected to expand in application as increasing amounts of newsprint must be deinked and recycled.

PERMISSIONS

All chapters in this book are published with permission under the Creative Commons Attribution Share Alike License or equivalent. Every chapter published in this book has been scrutinized by our experts. Their significance has been extensively debated. The topics covered herein carry significant information for a comprehensive understanding. They may even be implemented as practical applications or may be referred to as a beginning point for further studies.

We would like to thank the editorial team for lending their expertise to make the book truly unique. They have played a crucial role in the development of this book. Without their invaluable contributions this book wouldn't have been possible. They have made vital efforts to compile up to date information on the varied aspects of this subject to make this book a valuable addition to the collection of many professionals and students.

This book was conceptualized with the vision of imparting up-to-date and integrated information in this field. To ensure the same, a matchless editorial board was set up. Every individual on the board went through rigorous rounds of assessment to prove their worth. After which they invested a large part of their time researching and compiling the most relevant data for our readers.

The editorial board has been involved in producing this book since its inception. They have spent rigorous hours researching and exploring the diverse topics which have resulted in the successful publishing of this book. They have passed on their knowledge of decades through this book. To expedite this challenging task, the publisher supported the team at every step. A small team of assistant editors was also appointed to further simplify the editing procedure and attain best results for the readers.

Apart from the editorial board, the designing team has also invested a significant amount of their time in understanding the subject and creating the most relevant covers. They scrutinized every image to scout for the most suitable representation of the subject and create an appropriate cover for the book.

The publishing team has been an ardent support to the editorial, designing and production team. Their endless efforts to recruit the best for this project, has resulted in the accomplishment of this book. They are a veteran in the field of academics and their pool of knowledge is as vast as their experience in printing. Their expertise and guidance has proved useful at every step. Their uncompromising quality standards have made this book an exceptional effort. Their encouragement from time to time has been an inspiration for everyone.

The publisher and the editorial board hope that this book will prove to be a valuable piece of knowledge for students, practitioners and scholars across the globe.

INDEX